aus der Reihe:

Innovationen mit Mikrowellen und Licht

Forschungsberichte aus dem Ferdinand-Braun-Institut, Leibniz-Institut für Höchstfrequenztechnik

Band 70

Matthias M. Karow

Broad-Area Laser Bars for 1 kW-Emission.
Demonstrating increased efficiency and narrow far field

Herausgeber: Prof. Dr. Günther Tränkle

Ferdinand-Braun-Institut
Leibniz-Institut
für Höchstfrequenztechnik (FBH)
Gustav-Kirchhoff-Straße 4
12489 Berlin

Tel. +49.30.6392-2600
Fax +49.30.6392-2602

E-Mail fbh@fbh-berlin.de
Web www.fbh-berlin.de

Innovations with Microwaves and Light

Research Reports from the Ferdinand-Braun-Institut, Leibniz-Institut für Höchstfrequenztechnik

Preface of the Editors

Research-based ideas, developments, and concepts are the basis of scientific progress and competitiveness, expanding human knowledge and being expressed technologically as inventions. The resulting innovative products and services eventually find their way into public life.

Accordingly, the *"Research Reports from the Ferdinand-Braun-Institut, Leibniz-Institut für Höchstfrequenztechnik"* series compile the institute's latest research and developments. We would like to make our results broadly accessible and to stimulate further discussions, not least to enable as many of our developments as possible to enhance everyday life.

This work treats highly efficient high-brightness diode-laser bars delivering 1 kW optical power. They are employed industrially as pump sources for solid-state laser systems and increasingly in direct material-processing applications. In order to increase the electro-optical conversion efficiency and to improve beam quality, detailed investigations of the limiting physical mechanisms were carried out. The technological optimizations based on these insights are discussed in this dissertation and led to experimental record values in terms of efficiency and lateral divergence.

We wish you an informative and inspiring reading

Günther Tränkle

Prof. Dr. Günther Tränkle
Scientific Director

The Ferdinand-Braun-Institut

The Ferdinand-Braun-Institut researches electronic and optical components, modules and systems based on compound semiconductors. These devices are key enablers that address the needs of today's society in fields like communications, energy, health and mobility. Specifically, FBH develops light sources from the visible to the ultra-violet spectral range: high-power diode lasers with excellent beam quality, UV light sources and hybrid laser systems. Applications range from medical technology, high-precision metrology and sensors to optical communications in space and integrated quantum technology. In the field of microwaves, FBH develops high-efficiency multi-functional power amplifiers and millimeter wave frontends targeting energy-efficient mobile communications as well as car safety systems. In addition, compact atmospheric microwave plasma sources that operate with economic low-voltage drivers are fabricated for use in a variety of applications, such as the treatment of skin diseases.

The FBH is a competence center for III-V compound semiconductors and has a strong international reputation. FBH competence covers the full range of capabilities, from design to fabrication to device characterization.

In close cooperation with industry, its research results lead to cutting-edge products. The institute also successfully turns innovative product ideas into spin-off companies. Thus, working in strategic partnerships with industry, FBH assures Germany's technological excellence in microwave and optoelectronic research.

Broad-Area Laser Bars for 1 kW-Emission with High Efficiency and Narrow Far Field

vorgelegt von

Matthias M. Karow, MS, M.Sc.

ORCID: 0000-0003-1186-0084

an der Fakultät IV - Elektrotechnik und Informatik
der Technischen Universität Berlin
zur Erlangung des akademischen Grades

Doktor der Naturwissenschaften
- Dr. rer. nat. -

genehmigte Dissertation

Promotionsausschuss:

Vorsitzender:	Prof. Dr.-Ing. R. Thewes
Erstgutachter:	Prof. Dr. G. Tränkle
Zweitgutachter:	Prof. Dr. M. Kneissl
Drittgutachter:	Prof. S. Sweeney, PhD

Tag der wissenschaftlichen Aussprache: 27. April 2022

Berlin 2022

Bibliografische Information der Deutschen Nationalbibliothek
Die Deutsche Nationalbibliothek verzeichnet diese Publikation in der Deutschen Nationalbibliografie; detaillierte bibliographische Daten sind im Internet über http://dnb.d-nb.de abrufbar.
1. Aufl. - Göttingen: Cuvillier, 2022
Zugl.: Berlin, Technische Universität, Diss., 2022
u. d. T. Broad-Area Laser Bars for 1 kW-Emission with High Efficiency and Narrow Far Field

Nonnenstieg 8, 37075 Göttingen
Telefon: 0551-54724-0
Telefax: 0551-54724-21
www.cuvillier.de

1. Auflage, 2022
Gedruckt auf umweltfreundlichem, säurefreiem Papier aus nachhaltiger Forstwirtschaft.

ISBN 978-3-7369-7626-9
eISBN 978-3-7369-6626-0

Es ist ein Gefühl der Bewunderung, das uns gefangen nimmt, so oft wir erfahren, wie die Wirklichkeit einem logischen Gedankengebäude sich fügt. [Laß00]

Abstract

GaAs-based high-power diode-laser bars using broad-area emitters deliver optical powers P of typically several hundred watts. These commonly 1 cm-wide chips ("cm-bars") find widespread use in a range of industrial applications whereof, e.g. materials processing has generated more than $6 billion revenue in 2019. Here, laser bars are used either as pump sources, e.g. for high-brightness solid-state lasers, or their emission is routed directly to the work piece ("direct diode"). A major drawback of these bars – and of broad-area lasers in general – is the comparably poor beam quality in lateral direction. It limits their use in these applications requiring highest brightness.

This dissertation studies laser bars emitting $P = 1$ kW optical power (wavelength $\lambda = 940$ nm) held at room temperature and seeks to identify those physical mechanisms currently limiting electrical-to-optical conversion efficiency as well as lateral beam quality. In doing so, several diagnostic studies on bars with varied lateral-longitudinal design are carried out and the effect of technological measures for performance optimization is analyzed. Investigation of the lateral beam quality bases on a herein proposed and realized concept, in which the far field is studied for each individual bar emitter. This allows resolving how far-field profiles vary along the bar width and the magnitude with which these variations affect the overall bar far field.

The extreme-double asymmetric (EDAS) large optical cavity-concept was found suitable for the vertical structure as it affords the combination of low areal series resistance, high modal gain, and low optical loss $\alpha_{\mathrm{i}} = 0.35\ \mathrm{cm}^{-1}$. An EDAS-based benchmark design (length $L = 4\,\mathrm{mm}$, 37 emitters of width $w = 186\,\mu\mathrm{m}$) showed conversion efficiency $\eta = 61\%$ at the operation point. Lengthening the resonator to $L = 6$ mm increased the efficiency to $\eta = 63\%$ due to reduced series resistance, scaling with the cross-sectional area of the current path. The low optical loss α_{i} of the vertical structure was found crucial for this improved performance as it helped maintaining high slope efficiency $S \approx 1.1$ W/A. Based on extrapolation, devices with yet longer cavities do, however, require further reduced optical loss in order to show even higher efficiency.
An alternative path for reduced series resistance was pursued by increasing the lateral fill-factor (FF) from 69% to 87% at fixed resonator length $L = 4$ mm. Keeping chips as short as possible is desirable from a production cost point of view. Here, the devices even outperformed the series resistance-based estimation, as this design with $w = 1095\,\mu$m-wide emitters also showed reduced roll-over owing to improved gain extraction at high emission power. This effect was found most notable at elevated thermal condition where the slope efficiency at the operation point increased from $S \approx 0.6$ W/A (baseline) to ≈ 1 W/A.

These advancements led to the record room-temperature efficiency $\eta = 66\%$ at 1 kW. Combining the technologies studied herein promises $\eta \geq 70\%$ as feasible near-term goal.

The second half of this thesis is devoted to the far-field width of such efficiently-emitting cm-bars. It was found that one can group the governing physical mechanisms into those of non-thermal and those of thermal nature, whose separation succeeded when experiments were carried out at varied duty cycle in quasi-continuous wave (QCW) testing. Low duty cycle corresponding to a thermal resistance $R_{\mathrm{th}} \approx 0.02\,\mathrm{K/W}$ enabled studying one effect of the first category, namely mechanical stress. Soldering intrinsically strained bar chips, themselves comprising of layers with different thermal expansion coefficients, onto non-ideally planar submounts of yet another expansion coefficient leaves these mounted diode lasers in a bent shape. This mechanical deformation, commonly referred to as "bar smile", is known as detrimental but its effect and severity onto the lateral far-field pattern little studied. Despite ascription of beam-property worsening to either transverse-magnetic (TM) emission or polarization-axis rotation in literature, the experiments in this work suggest that none of the above dominates the herein observed far-field widening at increased smile but instead photoelastically-induced waveguides appear to shape the emission. A technological measure that reduced the propagation of external stress fields into the critical semiconductor layers was examined herein and enabled a lateral divergence as low as $\Theta_{95\%} = 8.8°$ (95%-power-content width) at 1 kW – a benchmark-setting value.
The studies into thermal effects were carried out at elevated duty cycle corresponding to a thermal resistance $R_{\mathrm{th}} \approx 0.05\,\mathrm{K/W}$ and thus at a heating level feasible in CW operation when advanced, but demonstrated, cooling technology is employed. It is known that in broad-area lasers, a bell-shaped lateral temperature profile builds up with increasing operation point and that it creates a lateral waveguide shaping and widening the emission. In this work, the effect of thermal emitter cross-talk is exploited as a means to flatten these lateral temperature profiles for weak lateral waveguiding resulting in halved far-field degradation. Structures making use of this beneficial emitter interaction fed 1 kW into an $\Theta_{95\%} = 10.8°$ angle at this elevated heating level. Extrapolating the measured behavior of emitters of different stripe width w to a proposed bar configuration projects $\Theta_{95\%} = 9.2°$ as feasible at this thermal resistance.

Kurzfassung

GaAs-basierte Hochleistungs-Diodenlaser-Barren mit Breitstreifen-Emittern liefern optische Ausgangsleistungen P von typischerweise einigen hundert Watt. Diese gewöhnlich 1 cm breiten Chips („cm-Barren“) finden weitverbreitete Verwendung in einem Spektrum von industriellen Anwendungen, wovon die Materialbearbeitung im Jahr 2019 mehr als 6 Milliarden $ Umsatz generierte. Hierbei werden Laserbarren entweder als Pumpquelle, z.B. für hochbrillante Festkörperlaser verwendet oder deren Emission wird direkt zum Werkstück geleitet („direct diode“). Ein großer Nachteil dieser Barren – wie von Breitstreifen-Lasern im Generellen – ist die vergleichsweise schlechte Strahlqualität in lateraler Richtung. Diese limitiert deren Verwendung in diesen Anwendungen, die höchste Brillanz erfordern.

Diese Dissertation untersucht Laserbarren mit $P = 1$ kW Leistung (Wellenlänge 940 nm) bei Raumtemperatur und ermittelt jene physikalischen Mechanismen, die die elektrisch-optische Konversionseffizienz η und die laterale Strahlqualität limitieren. Dafür werden diagnostische Untersuchungen an Barren mit variiertem longitudinal-lateralen Design durchgeführt und die Wirksamkeit von technologischen Maßnahmen zur Struktur-Verbesserung analysiert. Die Untersuchung zur lateralen Strahlqualität basiert auf einem hierin entwickelten Konzept, bei dem das Fernfeld eines jeden Barren-Emitters einzeln vermessen wird. Dies ermöglicht das Auflösen von Fernfeld-Variationen entlang der Barrenbreite und mit welchem Anteil diese das Gesamt-Barrenfernfeld beeinflussen.

Das Konzept der extrem-doppel-asymmetrischen (EDAS) großen optischen Kavität als Vertikalstruktur ermöglichte die Kombination aus geringem spezifischen elektrischen Widerstand, hohem modalen Gewinn und geringer optischer Verluste $\alpha_i = 0.35\ \text{cm}^{-1}$. Ein im Folgenden als Bezugswert herangezogenes Design mit EDAS-Vertikalstruktur (Länge $L = 4$ mm, 37 Emitter von $w = 186\ \mu$m Breite) zeigte eine Effizienz von $\eta = 61\%$ am Betriebspunkt. Durch Verlängerung des Resonators zu $L = 6$ mm erhöhte sich diese zu $\eta = 63\%$ durch verringerten Serienwiderstand, da dieser mit der Querschnittsfläche des Strompfades skaliert. Die geringen optischen Verluste α_i waren dabei besonders wichtig, da durch diese die Steigungseffizienz bei $S \approx 1.1$ W/A gehalten werden konnte. Extrapolationen zu noch längeren Kavitäten zeigten jedoch, dass zusätzlich verringerte optische Verluste nötig wären, um damit weiter verbesserte Effizienz zu erreichen.
Ein alternativer Weg zur Serienwiderstands-Verringerung wurde verfolgt, indem bei konstanter Länge $L = 4$ mm stattdessen der laterale Füllfaktor (FF) von 69% auf 87% erhöht wurde. In den Experimenten übertrafen diese Laser die auf der Serienwiderstands-Verringerung basierenden Schätzungen, da mittels der $w = 1095\ \mu$m breiten Emitter sich

außerdem das Überrollen reduzierte. Diese verbesserte Gewinn-Extraktion bei hohen Leistungen wurde besonders deutlich bei erhöhter Erwärmung, bei der sich die Steigungseffizienz von $S \approx 0.6\,\mathrm{W/A}$ (Bezugsdesign) auf $\approx 1\,\mathrm{W/A}$ verbesserte. Diese Verbesserungen ermöglichten eine Rekordeffizienz bei Raumtemperatur von $\eta = 66\%$ bei 1 kW.
Kombination der hierin untersuchten Technologien sollte $\eta \geq 70\%$ als Nahziel ermöglichen.

Die zweite Hälfte dieser Dissertation ist der Fernfeld-Breite solcher cm-Barren gewidmet. Die bestimmenden physikalischen Mechanismen wurden in solche von nicht-thermischer und solche von thermischer Natur unterteilt, wobei deren Separierung gelang, indem die Experimente bei variiertem Tastverhältnis des Quasi-Dauerstrich-Betriebes (QCW) durchgeführt wurden. Geringes Tastverhältnis, das einem thermischen Widerstand $R_{\mathrm{th}} \approx 0.02\,\mathrm{K/W}$ entspricht, ermöglichte es, den nicht-thermischen Effekt der mechanischen Verspannung zu untersuchen. Das Löten von Barren-Chips auf Wärmespreizer resuliert zumeist in mechanischer Deformation – auch als „Barren-Smile“ bezeichnet. Dieser ist als schädlicher Effekt bekannt, sein Einfluss auf das laterale Fernfeld jedoch wenig untersucht. Die Verschlechterung der Strahleigenschaften wird in der Literatur der transversal magnetischen (TM) Emission oder einer Polarisationsachsen-Drehung zugeschrieben. Im Gegensatz dazu suggerieren die Experimente dieser Arbeit, dass keiner der obengenannten Effekte die ermittelte Fernfeldverbreiterung mit wachsendem Smile dominiert, sondern dass photoelastisch induzierte Wellenleiter die Emission formen. Ein technologischer Ansatz, der die Propagation von externen Verspannungsfeldern in kritische Halbleiterschichten verringert, ermöglichte eine laterale Divergenz von $\Theta_{95\%} = 8.8°$ (95%-Leistungsinhalt) bei 1 kW – eine neue Referenz in diesem Feld.
Das Studium der thermischen Effekte wurde bei erhöhtem Tastverhältnis, einem thermischen Widerstand $R_{\mathrm{th}} \approx 0.05\,\mathrm{K/W}$ entsprechend, durchgeführt und damit bei einem Erwärmungslevel, das auch im CW-Betrieb unter Verwendung von fortschrittlicher Kühlungstechnologie realisierbar ist. Es ist bekannt, dass bei steigendem Betriebspunkt sich in Breitstreifen-Lasern ein glockenförmiges laterales Temperaturprofil entwickelt, das einen die Emission formenden und verbreiternden Wellenleiter verursacht. In dieser Arbeit wird nun die Auswirkung von thermischer Emitter-Wechselwirkung ausgenutzt, um diese Temperaturprofile für nur schwache laterale Wellenleitung zu glätten. Im Resultat reduzierte dies die Fernfeld-Verbreiterung um die Hälfte und Strukturen, die diesen Effekt ausnutzen, emittierten 1 kW in $\Theta_{95\%} = 10.8°$ auf diesem erhöhten Erwärmungslevel. Projektion des gemessenen Verhaltens von Emittern unterschiedlicher Breite w auf eine vorgeschlagene Barren-Konfiguration schätzt, dass $\Theta_{95\%} = 9.2°$ möglich ist.

Contents

List of Figures

List of Tables

List of Variables

Variable	Physical Quantity	Variable	Physical Quantity
A	Area	P	Optical power
α_i	Internal loss	P_diss	Dissipated power
c	Vacuum speed of light	$\Delta\phi$	Phase shift
C_th	Thermal capacitiy	p_sub	Sub-emitter pitch
c_th	Specific thermal capacity	$\dot{q}$	Heat flux density
d_s	Emitter spacing	$\vec{r}$	Position in space
d_th	Characteristic thermal length	$R_\mathrm{f}, R_\mathrm{r}$	Front / rear reflectivity
e	Elementary charge	ϱ	Mass density
$\vec{E}$	Electric field	ρ_a	Areal series resistance
ϵ	Permeability	R_s	Series resistance
ε_{kl}	Strain	R_th	Thermal resistance
η	Conversion efficiency	S	Slope efficiency
η_d	Differential efficiency	σ_{kl}	Mechanical stress
η_i	Internal efficiency	t	Time
f	Focal length	T	Temperature
f_rep	Repetition rate	T_0	Threshold-char. temp.
g_0	Material gain	T_1	Slope-characteristic temp.
Γ	Confinement factor	T_AZ	Active zone temperature
h	Planck's constant	T_HS	Heat sink temperature
h_chip	Chip thickness	Δt_p	Pulse length
i	Emitter number	θ	Far-field angle
I	Current	Θ	Far-field width
I_th	Threshold current	$\Delta\theta_\mathrm{res}$	Angular resolution
j_th	Threshold current density	τ_th	Thermal time constant
j_tr	Transparency current density	V	Voltage
L	Cavity length	V_0	Voltage offset
λ	Wavelength	V_d	Defect voltage
λ_th	Thermal conductivity	w	Stripe width
M	Magnification	w_ap	Bar aperture
n	Mode-order number	w_sub	Sub-emitter width
n_e	Number of emitters	x, y, z	Spatial coordinates
n_r	Refractive index	Δx_res	Spatial resolution
p	Optical power density	Z_th	Thermal impedance

List of Abbreviations

AlAs	-	aluminum arsenide
AZ	-	active zone
BAL	-	broad-area laser
BAR	-	broad-area laser bar
CCD	-	charge-coupled device
CCP	-	conduction-cooled package
CTE	-	coefficient of thermal expansion
CW	-	continuous wave
DC	-	duty cycle
DL	-	diode laser
DOP	-	degree of polarization
EF	-	emitter filter
FBH	-	Ferdinand-Braun-Institut, Leibniz-Institut für Höchstfrequenztechnik
FF	-	Fill-factor
FWHM	-	full-width at half-maximum
GaAs	-	gallium arsenide
GG	-	gain guiding
HPDL	-	high-power diode laser
IG	-	index guiding
IGT	-	index-guiding trench
InAs	-	indium arsenide
NA	-	numerical aperture
PL	-	photo-luminescence
QCW	-	quasi-continuous wave
ROP	-	rotated degree of polarization
RPL	-	relative power loss
TE	-	transverse-electric
TM	-	transverse-magnetic

1. Introduction

Diode lasers have turned into a ubiquitous component used in a wide range of modern-day applications. This is due to a number of desirable properties, including low cost per emitted power, high electrical-to-optical conversion efficiency, small footprint, wavelength versatility, and long lifetimes. They are applied in, for instance, fiber-based optical communication systems, in chemical and ranging sensors, medical apparatus, e.g. for surgery, as well as in industrial tools for materials processing [Bac07].
As one prominent diode-laser representative, this thesis studies GaAs-based high-power broad-area laser bars emitting in the 9xx nm wavelength range. Such bars typically come as 1 cm-wide chips, have several individual diode lasers (emitters) processed onto them, and thereby upscale the total emitted power to between several hundred and up to two thousand watts.
They are installed in many industrial lasers for materials processing techniques, the latter spanning from plastic marking to sheet metal cutting, named in the order of increasing required power density delivered to the work piece. Other emerging processing techniques include high-precision micro-fabrication and additive metal printing. Industrial lasers can emit powers of the 100 kW-class and depend on high-power diode lasers (HPDL) – such as the herein studied high-power laser bars – as a key component. This workhorse of industrial laser technology is typically designed from broad-area lasers (BAL) and excels as the most efficient device to convert electrical into optical power. The emission of HPDLs is either routed directly to the processing site ("direct diode") [Ried18] or used to pump fiber and solid-state lasers, such as disk lasers [McD20]. Owing to the comparably poor beam quality (low brightness) of HPDLs, direct applications are limited to those requiring lower power density, e.g. hardening and marking, although consistent performance improvements increasingly enable their usage for even welding and brazing. Fiber

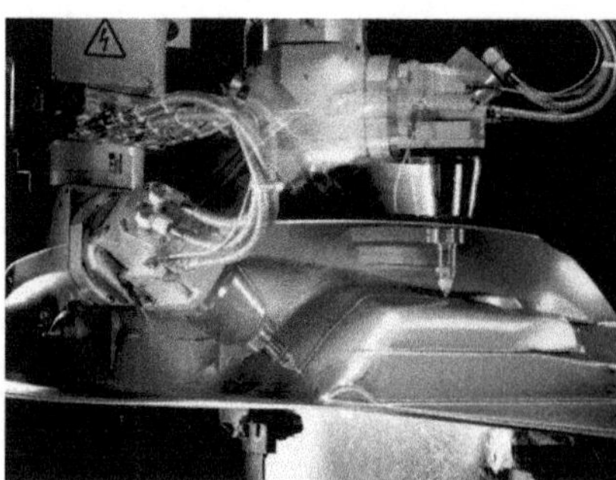

Figure 1.1: Laser-based metal cutting of a car body component ©TRUMPF Group

and solid-state lasers, on the other hand, offer higher brightness and are thus capable of, for instance, sheet metal cutting and deep-penetration welding. Here, HPDLs pump the used gain medium (e.g. Yb-doped fiber or crystal, respectively) at one of its absorption maxima and so provide the gain required for lasing. In effect, fiber and solid-state lasers enhance the input brightness of HPDLs, e.g. by a factor $> 20,000\times$ in a state-of-the-art fiber laser [Kan18]. Improvements in HPDL brightness generally allow further scaling in output power from the pumped lasers and bettered overall efficiency.

High-power laser bars – the matter of this dissertation – benefit from the advantages of their individual broad-area lasers, such as high conversion efficiency, but they also inherit their major drawback – the comparably poor beam quality. What is worse, bars were found to provide inferior beam quality in comparison to equivalent single-emitter lasers operated at the same per-emitter-power [Cru17]. This may be a pure ensemble effect but additional causes seem likely and need to be identified in order to eventually find solutions for improved bar beam quality.

It is thus at the heart of this thesis to study which physical mechanisms limit and which design optimizations enhance the performance of high-power diode-laser bars. In doing so, this work represents a contribution to the global optimization and design efforts that have allowed the brightness of HPDLs to increase by a factor of ten every eight years for the last 35 years [Kan18].

The remainder of this chapter firstly assesses the global laser market to showcase the economic relevance of the field, then analyzes the state of the art by comparing benchmark lasers, and finally outlines the structure of this thesis.

1.1. The global laser market

The world-wide laser market is predicted to amount to nearly \$17 billion in 2020 revenue, reached by a steady annual growth of on average 12% p.a. since 2015, cf. Fig. 1.2(a). Exceptional growth rates of nearly 20% were observed in 2016 and 2017. About 42% of the lasers sold use diode technology as the core component while the remaining portion in many cases makes use of diode lasers as an auxiliary device, e.g. as pump laser. Grouped into market segments, about \$6 billion were spent in 2019 on lasers for materials processing and lithography applications. This constitutes the largest market share of 40%, followed by the communications and optical storage sector (27%), cf. Fig. 1.2(b). The industrial lasers sold in the dominating market segment serve as tools for a variety of materials processing techniques, whose relative frequency is illustrated in Figure 1.2(c), and of which cutting (35%) is the by far the most important one. Three further applications, namely welding/brazing, marking, and semiconductor/display show similar shares of each

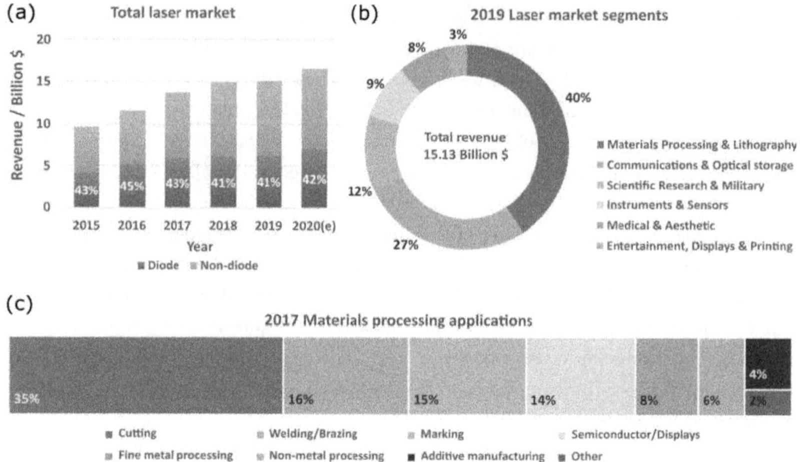

Figure 1.2: Metrics of the global laser market. (a) Annual revenue of the total laser market split into diode and non-diode contributions. (b) Market segment shares of the 2019 laser sales. (c) Applications of lasers sold in the materials processing segment in 2017. Data taken from [Hol18a] and [Gef20].

about 15%.

1.2. State of the art

The following firstly gives an overview of current industrial laser systems along with their specifications and secondly reviews recent notable publications on high-power laser bar performance.

Which materials processing technique a laser system is suitable for is determined by the emitted power P and the corresponding beam quality, quantified for instance via the *beam parameter product* BPP as introduced in Section 2.2. Combined, the *linear brightness* $B_{\text{lin}} = P/\text{BPP}$ expresses the power density of the emitted beam and it is among the most important parameters to characterize a beam. Table 1.1 lists commercial laser systems with top performance from a variety of vendors. They are based on four different technologies, namely fiber, disk, CO_2, and direct-diode lasers, of which disk and many direct-diode lasers incorporate high-power laser bars. The highest linear brightness B_{lin} is provided by fiber lasers with values up to 5.3 W/mm mrad, enabled by near-diffraction-limited beam quality BPP = 0.38 mm mrad, and are thus suitable for the most demanding

B_{lin} #	P / kW	BPP †	η / %	λ / nm	Type	Vendor / Product
5.3	2	0.38	32	1075	fiber	Trumpf TruFiber 2000
3.5	60	17	45	1070	fiber	IPG YLS-60000
3.0	12	4.0	37	1030	disk	Trumpf TruDisk 12001
2.3	8	3.5	8	1060	CO_2	Coherent DC 080
1.6	10	6.4	9	1060	CO_2	Trumpf TruFlow 10000
0.28	11	40	35	900-1080	direct DL	Laserline LDF series
0.15	4.5	30	40	920-970	direct DL	Trumpf TruDiode 4506
0.15	1.5	10	33	976	direct DL	IPG DLR-976-1500
0.13	4	30	31	900-1100	direct DL	Coherent HL DL4000HQ

Table 1.1: Commercial laser systems for high-power material processing applications ranked with descending brightness (product with highest available brightness selected per vendor and type), units # kW / mm mrad, † mm mrad. Specifications taken from vendors' online product catalogs as of January 2021, private communication with vendor, and [Ried18].

Publication	Year	P / kW	η / %	Operation	T / °C	FF / %	L / mm
[Sch07]	2007	1.1#	34	QCW	15	44	4
[Li08]	2008	1.0	56	CW	8	83	5
[Kna11]	2011	0.94	55	CW	20	77	5
[Fre15]	2015	1.98	57	QCW	-73	69	4
[Fre16]	2016	1.0	70	QCW	-70	69	4
[McD20]	2020	0.24§	59	CW	29	60	4
This work	2020	1.0	66	QCW	25	87	4
[Cru21]	2021	0.8	61	CW	15	87	4

Table 1.2: Notable publications on kW-class cm-bars, here focusing on emitted power P, the efficiency η at this operation point, and the used heat sink temperature T. All bars emit at 940 nm, except #at 980 nm. §state of the art in large-scale high-reliability production frame

processing applications. Disk lasers, here one using a Yb:YAG gain medium pumped by high-power laser bars, show brightness values as high as 3 W/mm mrad. The CO_2 laser is a mature technology that is steadily being replaced by the competing technologies not only due to its poor conversion efficiency. Direct-diode lasers excel in reduced system complexity as no optical to optical conversion step is required and make use of wavelength and polarization multiplexing for power scaling. Their brightness still falls behind the competing platforms which is due to their comparably poor beam quality with a BPP = 40 mm mrad in case of the highest brightness product listed here (0.28 W/mm mrad).

The object of study in this thesis is the 1 cm-wide high-power broad-area laser bar. Industrial large-scale fabrication with high reliability reaches power levels of [200:300] W as published in, for instance, [McD20] where 240 W-emitting bars for pump applications

Publication	Year	P / kW	$\Theta_{l,\,95\%}$ / °	η / %	R_{th} / KW^{-1}	Operation	T / °C	FF / %
[An15]	2015	0.29	8.3	58	0.21	CW	25	50
[Hei18]	2018	0.27	8.5	54	0.24	CW	29	60
[Str18]	2018	0.65	10.8	64	0.054	CW	15	70
This work	2020	1.0	8.8	61	0.02	QCW	25	69
This work	2020	1.0	10.8	54	0.05	QCW	25	69

Table 1.3: Notable publications on kW-class cm-bars, here focusing on emitted power P and lateral divergence angle $\Theta_{l,\,95\%}$ at this operation point. All bars emit at 940 nm.

are presented. In an effort to minimize the cost per emitted power (from < \$2 /W in aforementioned publication) ever increasing operation points have been targeted with 1 kW per bar as a landmark. The first demonstration that has reached this power level was published in 2007 and made use of a 4 mm-long resonator and 44% fill-factor (FF, that is the fraction of bar width occupied by active emitters) [Sch07]. The bar was driven in quasi-continuous wave (QCW) mode with an efficiency of 35%, cf. Tab. 1.2. After this milestone had been reached, the conversion efficiency needed to be improved, which suffered from the high series resistance $R_s = 1.3\,\mathrm{m\Omega}$. One year later, the first continuous wave (CW) 1 kW-demonstration was reported with increased 83% FF and 5 mm cavity length [Li08] using high-heat-removal-capacity packages with a low thermal resistance $R_{th} = 0.08\,\mathrm{K/W}$. From thereon more and more prototypes for this power level were developed. Another milestone in kW-class bar research was achieved at cryogenic temperatures −73°C enabling nearly 2 kW emitted power [Fre15], with record 70% efficiency at 1 kW [Fre16]. This thesis focuses on "room-temperature" operation as an industrial boundary condition instead and its design studies have let the 1 kW-efficiency increase to 66% in QCW-mode [Kar20]. Recent activities in which bars developed within this work were mounted in low-thermal-resistance packages reached 61% at 0.8 kW in CW mode [Cru21].

Aside from the desire for ever higher conversion efficiency, the beam quality of the emitted light also determines the merit of a bar design. An overview of notable publications on this matter is given in Table 1.3. Industry developments feed about 300 W in a lateral angle $\Theta_{l,\,95\%}$ of about 8.5° in CW mode [An15, Hei18]. The heating in the devices has been identified as a major determinant of the lateral divergence, as will be elaborated on in Section 2.2, and it is quantified via the thermal resistance R_{th}, also listed in the table. An industrial prototype increased the emitted power to 0.65 kW at a divergence of 10.8°, also tested in CW mode but using high-capacity coolers with low thermal resistance $R_{th} = 0.054\,\mathrm{K/W}$ [Str18]. Developments in the frame of this work allowed feeding 1 kW into an angle of 8.8° in QCW mode (using $R_{th} = 0.02\,\mathrm{K/W}$) [Kar20].

1.3. Structure of this work

This thesis aims to study which physical mechanisms affect and which technological approaches improve the performance of 1 kW-emitting broad-area laser bars, whereby the focus is laid on the conversion efficiency η and the lateral divergence angle $\Theta_{l,\,95\%}$ at the operation point.

In doing so, Chapter 2 provides the theoretical base of this work by firstly describing the design and working principle of broad-area laser bars and their comprising component – broad-area lasers (BAL). The work then summarizes what has been found to this date about the mechanisms governing the beam quality of BALs and thus provides insight on which design studies and the use of which technologies appear worthwhile for performance progression. Chapter 3 outlines the fabrication steps of the devices processed for this work and then details the characterization techniques and setups with a focus on the emitter-resolved beam-quality test station conceived and realized in the course of this thesis. The experimental results are divided into design studies seeking to increase the conversion efficiency, cf. Chapter 4, and those to lower the lateral divergence, cf. Chapter 5, while building on the findings from Chapter 4. Conclusions on the physical insights gained and the performance progresses achieved are complemented by a prospect to this work in Chapter 6.

2. Theoretical Background

This chapter lays the theoretical basis of this work. While it does not aim to span across all topics associated with diode-laser bars or to provide a comprehensive treatment, it shall highlight important equations and name relevant sources for further study.

2.1. Concept and physics of broad-area laser bars

Diode-laser bars are comprised of several individual diode lasers (emitters) processed onto one macroscopic chip of semiconductor material, cf. Fig. 2.1. The resultant incoherent superposition of these emitter's radiation aims to upscale the emission power per device. At the same time, this monolithic integration of several light sources reduces manufacturing costs (in $/W-optical-power) and shrinks total device footprint as compared to individually manufactured single-emitter lasers. The operation voltage for the latter and for bars of the same emitter type are comparable as the individual bar emitters are circuited in parallel. A common quantity used to characterize bar designs is the (lateral) fill-factor (FF). It reflects the fraction of the total bar width (of the distance between the edges of the two outmost emitters, to be exact: "the aperture") that is comprised of active emitters. For a number n_e of w-wide stripes with a spacing d_s in-between adjacent emitters[1] the FF reads

$$\mathrm{FF} = \frac{n_\mathrm{e} w}{w_\mathrm{aperture}} = \frac{n_\mathrm{e} w}{(w + d_\mathrm{s})(n_\mathrm{e} - 1) + w}.$$

The basic operation principle and the underlying physics does not differ between laser bars and single-emitter diode laser.

A plethora of literature covering semiconductor devices and diode lasers in particular is available. A few introductory books shall be mentioned for different foci of interest. [Sze07] treats a wide range of semiconductor devices as well as essential building blocks and several opto-electronic devices and may thus be a suitable starting point. A focus on the physics of diode lasers is laid on in [Col95] and [Chu12] deriving essential underlying equations, while [Epp13] is rather engineering-oriented and considers real-world laser designs and their design parameters. Diode lasers for high-power applications are covered by [Die00], also including chapters on fabrication, and by [Bac07] further detailing packaging and beam-combining systems.

Diode-laser bars for high-power applications typically make use of broad-area lasers (BAL) which are of the edge-emitting type. This is, a Fabry-Pérot resonator is formed by two facets cleaved perpendicular to the epitaxial growth direction (longitudinal modal

[1] The quantity pitch p is often used in literature and can be computed via $p = w + d_\mathrm{s}$.

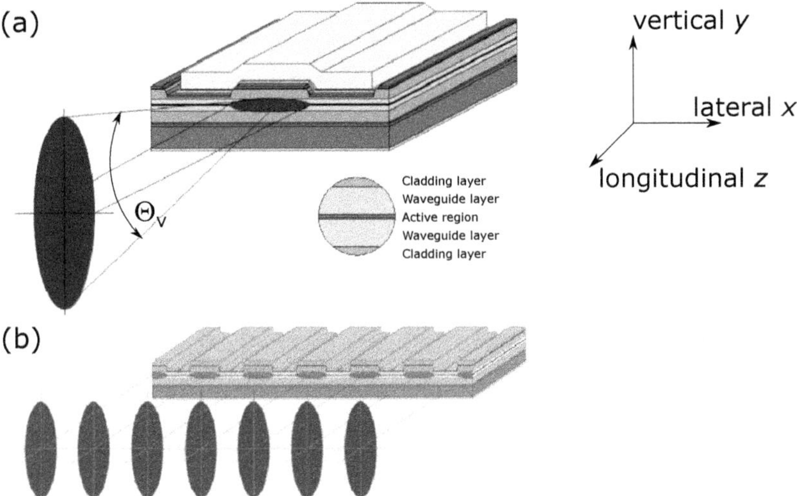

Figure 2.1: Illustration of broad-area lasers (BAL) in the further used coordinate system (a) Single-emitter laser with exemplary emission pattern (b) Laser bar comprised of several emitters processed on-chip. Not to scale, taken from [Erb00] ©Springer-Verlag

confinement). The resonator measures typically $1\,\text{mm} \leq L \leq 6\,\text{mm}$ in length and the laser light is extracted through of the facets exhibiting low reflectivity $R_\text{f} \leq 10\%$. Pumping is realized via charge carrier injection through a lateral-longitudinal current-path cross-section of width w, i.e. the stripe width, and is facilitated by a p-i-n-junction doping profile subject to forward bias. In case of BALs, w measures orders of magnitude above the emission wavelength $w \gg \lambda$ with typical values $50\,\mu\text{m} \leq w \leq 200\,\mu\text{m}$. Vertical confinement is nowadays realized via SCH-designs (seperate confinement heterostructure) as opposed to earlier double-heterostructure devices, expressing that modal and charge carrier confinement are realized by individually acting layer sequences. For emission in the 9xx nm-range, as targeted herein, the vertical waveguide is commonly grown from the $\text{Al}_x\text{Ga}_{1-x}\text{As}$ semiconductor compound whereby core and cladding layers (refractive index profile formed via molar fraction x and layer thicknesses) shape the vertical mode such that the desired mode width, position, confinement factor Γ and overlap with doped regions (one contributor to the internal loss α_i) is realized. Charge carrier confinement to reach the high carrier densities necessary for stimulated emission is commonly realized via strained $\text{In}_x\text{Ga}_{1-x}\text{As}$ quantum wells offering high material gain g_0 due to their narrow

charge-carrier energy distribution. The incorporated strain and the resultant energetic separation of heavy- and light-hole band further facilitate polarized emission; in the given case of compressive in-plane strain being of the transverse-electric (TE) type.
The mathematical description of the emission and recombination processes inside a laser diode [Col95, Chu12] find the threshold current I_{th} as

$$I_{\mathrm{th}} = wLj_{\mathrm{tr}} \cdot \exp\left(\frac{\alpha_{\mathrm{i}} + \alpha_{\mathrm{m}}}{\Gamma g_0}\right), \qquad \alpha_{\mathrm{m}} = -\frac{1}{2L}\ln\left(R_{\mathrm{f}}R_{\mathrm{r}}\right), \tag{2.1}$$

with the transparency current density j_{tr}, the mirror loss α_{m}, and the front and rear facet reflectivity values R_{f} and R_{r}, respectively. The emitted power P above threshold reads

$$P = \frac{hc}{e\lambda} \cdot \frac{\eta_{\mathrm{i}}}{1 + \alpha_{\mathrm{i}}/\alpha_{\mathrm{m}}} \cdot (I - I_{\mathrm{th}}),$$

with η_{i} the internal efficiency – characterizing the laser material – as the fraction of charge carriers that undergo stimulated recombination, as well as Planck's constant h, c the vacuum speed of light, and e the elementary charge. The slope efficiency S is defined as the derivative of the emission power

$$S = \frac{\mathrm{d}P}{\mathrm{d}I} = \frac{hc}{e\lambda} \cdot \frac{\eta_{\mathrm{i}}}{1 + \alpha_{\mathrm{i}}/\alpha_{\mathrm{m}}} \tag{2.2}$$

and reflects, reduced to the differential efficiency η_{d},

$$\eta_{\mathrm{d}} = \frac{\eta_{\mathrm{i}}}{1 + \alpha_{\mathrm{i}}/\alpha_{\mathrm{m}}},$$

the fraction of injected charge carriers above threshold that are converted to and emitted as lasing photons. The $I - V$ characteristic of diodes can be approximated to first order using

$$V(I) = V_0 + R_{\mathrm{s}} \cdot I,$$

with V_0 the voltage offset required to reach flatband conditions and R_{s} the equivalent series resistance the charge carriers experience whilst drifting and diffusing through the laser structure. From the equations above, the electrical-to-optical conversion efficiency η is derived as

$$\eta = \frac{P}{IV} = \eta_{\mathrm{i}} \cdot \frac{1}{1 + \alpha_{\mathrm{i}}/\alpha_{\mathrm{m}}} \cdot \frac{hc}{e\lambda(V_0 + R_{\mathrm{s}}I)} \cdot (1 - I_{\mathrm{th}}/I).$$

This gives a first impression on how to improve the conversion efficiency of a certain laser structure. While a high value η_{i} is required, α_{i}, R_{s}, and I_{th} need to be minimized. The particularity of this are the mutual dependencies between the parameters which compli-

cate efforts for efficiency improvements.
Non-radiative recombination processes, such as Shockley-Read-Hall and Auger recombination, as well as thermalization processes at band structure discontinuities or upon free-carrier-absorption heat up the structure. The dissipated power P_{diss} can be expressed as a function of the electro-optical efficiency η and the optical power P. Heat removal via a cooling architecture with the thermal resistance $R_{\text{th}} = \mathrm{d}T/\mathrm{d}P_{\text{diss}}$ allows the device to heat up by the temperature increase[2] ΔT_{AZ}.

$$P_{\text{diss}} = IV - P = (1/\eta - 1)P \quad \Rightarrow \quad \Delta T_{\text{AZ}} = R_{\text{th}}(1/\eta - 1)P \tag{2.3}$$

The heating itself has manifold implications on the laser performance which essentially can be traced back to a change in lattice constants (and the corresponding band diagram, i.e. band gap energy E_{g} and refractive index n_{r}) and altered, Fermi-Dirac-determined, occupation of the latter. This affects, for instance, the gain distribution, the waveguide action, the carrier transport, and the relative strength of the different recombination channels. A simple but useful and commonly applied way of quantifying the impact of temperature changes onto laser performance defines the phenomenological characteristic temperatures T_0 and T_1, which describe the changes in threshold current and slope efficiency, respectively.

$$I_{\text{th}}(\Delta T) = I_{\text{th},0} \cdot \exp(\Delta T/T_0), \qquad S(\Delta T) = S_0 \cdot \exp(-\Delta T/T_1) \tag{2.4}$$

Here, $I_{\text{th},0}$ and S_0 denote the values from Eqs. 2.1 and 2.2 at a chosen baseline temperature. While the threshold increases with temperature, the slope efficiency decreases.

2.2. Definition and governing effects of beam quality in broad-area lasers

The suitability of BALs for specific applications, such as material processing or laser pumping, crucially depends on the power density one can reach when focusing the emitted light. This focusability is a fundamental property of a laser beam and is commonly referred to as its beam quality. It cannot be improved via beam shaping optics and is, in the best case of diffraction-limited optical elements, conserved upon propagation along an optical path. This is why efforts to reach high power density need to start right at the light sources and their emission process.
A widespread way to quantify beam quality is the use of the *beam parameter product* BPP that is composed of the beam width w_0 at its smallest extent (beam waist) and its full

[2]Here referring to the temperature of the active zone (AZ) as a crucial determinant of laser performance, e.g. via its provided gain distribution.

angle Θ_∞ at infinite distance.

$$\mathrm{BPP} = \frac{1}{4}\left(w_0 \cdot \Theta_\infty\right)$$

Its lower natural limit is given by a single mode, fully coherent Gaussian beam. This diffraction limit $\mathrm{BPP_{dl}}$ depends on the wavelength via

$$\mathrm{BPP_{dl}}(\lambda) = \lambda/\pi, \qquad \mathrm{BPP_{dl}}(940\,\mathrm{nm}) = 0.3\,\mathrm{mm\ mrad}.$$

This absolute measure for beam quality is complemented by the *beam propagation ratio*

$$\mathrm{M}^2 = \frac{\mathrm{BPP}}{\mathrm{BPP_{dl}}} = \frac{\mathrm{BPP}}{\lambda/\pi}$$

that relates the absolute beam quality to its lowest achievable value at this wavelength and thus expresses the space left for beam quality improvement. A definition of an actual power density is given by the *radiance*

$$B = \frac{P}{A\,\Omega} = \frac{P}{4A \cdot \arcsin(\sin(\Theta_x/2)\sin(\Theta_y/2))},$$

relating the emitted power P to the emission area A shining into the solid angle Ω. Its second expression is valid for a bi-axial emission pattern as typical for BAL with the two angles Θ_i, $i = x, y$ along the axes x and y [Wes13]. For a device whose beam quality varies only along one axis x, e.g. as a function of operation point, the dependence of the *linear radiance* ("brightness") may be of interest [Bac07].

$$B_{\mathrm{lin}} = \frac{P}{\mathrm{BPP}_x}$$

A topic of controversy by itself is the width definition of beam profiles [Sie98, Eic04]. In short, there is no absolute definition. The full-width-at-half-maximum, the variance, and the width at $1/e^2$ of the maximum intensity are all used in certain fields and applications. The latter definitions are founded in the mathematical properties of a Gaussian distribution and barely mirror the characteristic of beams used in high-power applications. In practice, emission profiles often exhibit irregular multi-mode shapes, such that a knife-edge-power-content definition is commonly chosen. The 95%-width $\Delta x_{95\%}$ encloses 95% of the integrated power of a beam profile $p(x)$ by leaving out 2.5% at each tail; it is a common industry standard.

$$\Delta x_{95\%} := x_2 - x_1, \quad \text{with} \quad \int_{-\infty}^{x_1} p(x)\mathrm{d}x = \int_{x_2}^{+\infty} p(x)\mathrm{d}x = 0.025 \cdot \int_{-\infty}^{+\infty} p(x)\mathrm{d}x$$

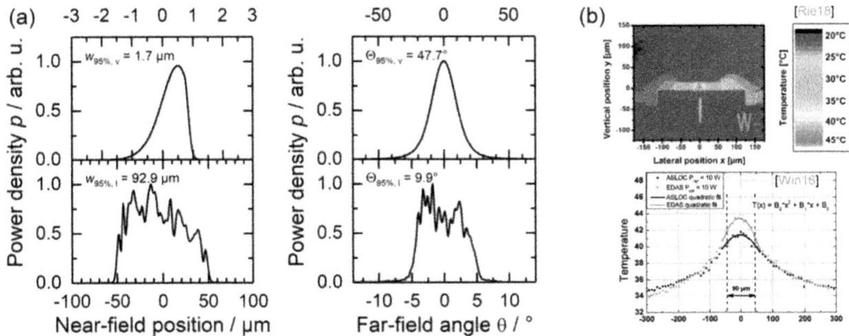

Figure 2.2: Beam quality in BALs (a) Comparison of vertical (top, simulation) and lateral (bottom, exemplary measurement of 90 μm-BAL at P = 4 W) beam quality with the corresponding 95%-widths. (b) Exemplary illustration of the vertical-lateral temperature distribution (chip in green and red) building up upon operation recorded via thermography, along with the lateral temperature profile ("thermal waveguide") extracted from such measurements. Taken from [Rie18, Win16] ©AIP Publishing and SPIE.

This definition will be used herein, e.g. for the near-field width $w_{95\%}$ and the far-field width $\Theta_{95\%}$.

The beam quality of BALs now exhibits a strong asymmetry in vertical and lateral direction. The emission in the vertical direction is designed to be at the fundamental mode and approximates a Gaussian shape. Figure 2.2(a) depicts exemplary near- and far-field profiles for the two principal axes x and y. Although comparably divergent ("fast axis", $\Theta_{95\%} = 47.7°$), the beam reveals excellent focusability in the vertical direction due to its high beam quality BPP= 0.35 mm mrad corresponding to $\mathrm{M}^2 = 1.2$ and thus close to the diffraction limit. The lateral ("slow") axis, in contrast, is the weak spot of BALs as here the emission is an incoherent superposition of several lateral modes. The profiles typically are of top-hat shape with irregular fluctuations and are further strongly dependent on the operation point. The exemplary profile recorded from a 90 μm-BAL operated at 4 W output power shows BPP= 4.0 mm mrad and $\mathrm{M}^2 = 14$ indicating significantly worse beam quality.

The following names the cardinal determinants of the beam quality of BALs and lists proposed measures for its optimization. Grouped into two categories one can distinguish causes related to the provided modal gain and those to the modal confinement of a lateral waveguide[3].

[3] which may also be perceived as the two contributions of one complex refractive index, split into imaginary and real parts, respectively

By design, BALs amplify a number of lateral modes owing to the wide injection region ($w \gg \lambda$), even at the lowest operation point, prohibiting diffraction-limited emission. Current spreading between contact layer and active zone further widens the region that provides gain – by $\Delta x_{\mathrm{spread}} \leq 10\mu\mathrm{m}$ [Pip13a] – and promotes high-order low-beam-quality modes. Lateral carrier diffusion and consequent accumulation act in similar fashion. Spatial hole burning – a result of gain extraction by low-order modes and insufficient carrier replenishment – is yet another contributor that causes the relative contributions of high-order modes to the total emitted power to increase. The community has proposed and/or realized the following methods to mitigate these factors and to improve beam quality:

- Deep implantation:

  Restriction of the lateral-vertical current path by highly resistive implanted regions beneath the stripe edges from contact layer to active zone [Win15, Mar19]

- Buried-mesa two-step epitaxy

  Etching and re-growth of a buried-mesa containing vertical waveguide and active zone for suppression of lateral carrier spreading [Del17]

- Series resistance tailoring:

  Series resistance reduction of the confining layers to facilitate carrier re-distribution and mitigate spatial hole burning [Joy82, Zeg19]

Aside from mode amplification, the presence of a lateral refractive index profile may guide a number of high-order modes, depending on the confining action of this waveguide. It was proposed to form a lateral structure using

- Resonant anti-guiding:

  Insertion of high-refractive-index regions beneath stripe edges coupling to high-order modes thereby reducing their modal gain [Wen13a]

The effect most frequently studied and in need of mitigation is the formation of a lateral temperature profile upon operation ("thermal waveguide", "thermal lens"), cf. [Bai11, Cru12, Win14, Bac17, Rie18]. The inhomogeneous generation of heat and its inhomogeneous conduction towards the heatsink leads to a confining lateral temperature – and thus refractive index [Ska03] – profile. An exemplary illustration of the vertical-lateral temperature distribution of an operated BAL is depicted in Figure 2.2(b). Not only does this effect allow for the support of modes with increasing mode number (and steadily decreasing beam quality), but it shapes the low-order modes in such a way as to also worsen their beam quality. It is seen as the key cause of beam-quality degradation at increasing operation power. Approaches to diminish this *thermal waveguide* seek to flatten any lateral temperature profile and include [Kar20]

- increased efficiency η at operation point for less heating ΔT_{AZ}, cf. Eq. 2.3 [Kar17a]
- reduced chip temperature ΔT_{AZ} via low-thermal-resistance R_{th} mounting [Str18, Kit15]
- tailored cooling architecture, e.g. using pedestals [Sun13]
- adjusted vertical layer structure for suitable thermal conductivity around dissipating volume [Liu17, Rie18]
- specific emitter design for optimized heat-source distribution
- flattened temperature profile by exploiting thermal emitter cross-talk in bars.

Of these measures this thesis makes use of deep implantation, increased efficiency, reduced thermal resistance, optimized heat-source distribution, and thermal emitter cross-talk in order to design high-power laser bars with low lateral emission divergence.

2.3. Strained diode-laser chips

Unintended stress in diode-laser chips is held accountable for a number of detrimental effects observed and is thus extensively studied [Tom06, Bie07, Wes08, Hem12]. It was reported to, for instance, compromise the high degree of polarization

$$\mathrm{DOP} := \frac{P_{\mathrm{TE}}}{P_{\mathrm{TE}} + P_{\mathrm{TM}}}$$

required in any laser system relying on polarization-multiplexing or to impair device reliability.

A number of factors is accountable for the strain field building up in diode-laser chips. Following [Hol18], one can coarsely group them into:

- Processing-related: epitaxial growth of semiconductor layers of different bulk lattice constants, thin-film deposited layers (e.g. metallization, passivation) of different coefficients of thermal expansion (CTE), mounting on non-ideally planar submounts with yet another CTE[4,5]
- Geometrical: shape and position of geometrical features, such as ridges, trenches, mesas, and openings, at which stress was found to accumulate, cf. Fig. 2.3(a) [Col93,

[4]For this reason, mounting onto expansion-matched submounts made of, e.g. CuW, is increasingly preferred over direct mounting onto Cu-heatsinks, at the cost of increased thermal and electrical resistance.

[5]Note that the device cools down after processing from the temperature at which the respective process step needs to be carried out and that thereby the different CTEs of the layers cause residual strain.

Cas04]

- Device temperature: heating of the device upon operation changes magnitude and distribution of the strain field, due to the different incorporated CTEs

The linear elasticity model described by Hook's law [Hoo78, Bar10] relates the strain ε_{mn} experienced by a material of the elasticity tensor c_{klmn} to the stress σ_{kl} it is subjected to via[6]

$$\sigma_{kl} = c_{klmn}\varepsilon_{mn}.$$

Changes in lattice constant described by the strain field ε_{mn} then affect the propagation of light and alters the permeability ϵ (impermeability ϵ^{-1}) from the unstrained case. This *photo-elastic effect* can be described via

$$(\Delta\epsilon^{-1})_{kl} = p_{klmn}\varepsilon_{mn},$$

with p_{klmn} the photo-elastic tensor [Ada83, Nye85].
A spatially varying strain field $\varepsilon_{mn}(\vec{r})$ of general nature then has three implications on the refractive index $n_{\mathrm{r}} = \sqrt{\epsilon}$ of the exposed medium:

- **Waveguiding $n_{\mathrm{r}} = n_{\mathrm{r}}(\vec{r})$**

The refractive index becomes a spatially varying function. In the case of a suitable distribution, this can lead to optical guiding or anti-guiding effects [Bud00]. [Kir79], for instance, isolated photo-elastically-induced lateral guiding in $\mathrm{Al}_x\mathrm{Ga}_{1-x}\mathrm{As}$ hetero-structures originating from a process window in the oxide cap. Their theoretical and experimental analysis found confinement of up to $\mathcal{O}(\Delta n_{\mathrm{r}}) = 10^{-2}$, increasing with the stripe width.

- **Birefringence $n_{\mathrm{r}} = n_{\mathrm{r}}(\vec{e}_{\vec{E}})$**

When subject to shear strain ε_{mn}, $m \neq n$, the refractive index of GaAs becomes a function of the plane of polarization $\vec{e}_{\vec{E}}$ the propagating light exhibits [Hig69]. This anisotropy causes a rotation of the polarization axis upon propagation and can give rise to TM-contributions stemming from an initially TE-polarized beam.

Changes in the refractive index are necessarily linked to changes in the band structure. Important for the discussion in this thesis is the energetic splitting between heavy- and light-hole valence band as dependent on normal strain:

- **Polarization-dependent gain $g_0 = g_0(\vec{e}_{\vec{E}})$**

[6] using Einstein notation

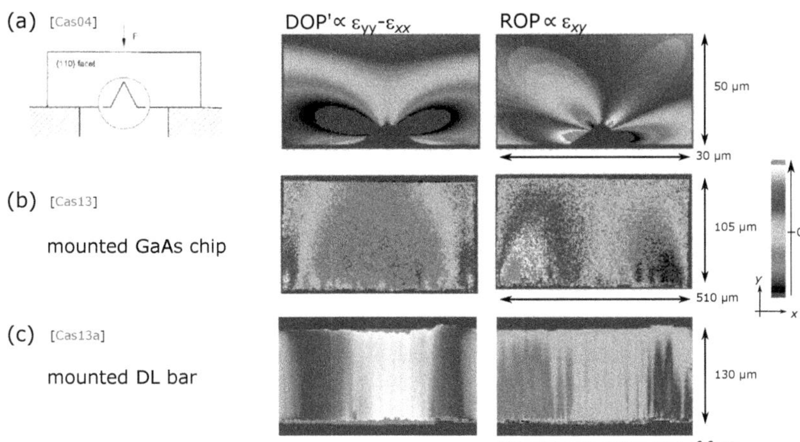

Figure 2.3: Normal and shear strain fields in stressed chips deduced from polarization- and spatially-resolved photo-luminescence experiments. (a) An etched v-groove accumulates stress resultant from an external downward force F [Cas04]. (b) Strain fields resulting from mounting of a GaAs chip of single-emitter dimensions onto a ceramic carrier (interface at bottom) [Cas13]. (c) Results from a diode-laser bar mounted onto a Cu submount (interface at bottom) [Cas13a]. In these publications, the definition DOP' $= (P_{\mathrm{TE}} - P_{\mathrm{TM}})/(P_{\mathrm{TE}} + P_{\mathrm{TM}})$ is used. ©OSA

Increasing compressive in-plane strain $-\varepsilon_{xx}$ raises the energetic splitting between heavy- (facilitating TE-gain) and light-hole (TM-gain) valence band, thereby alters the occupation of the two, and then favors optical gain for in-plane TE emission [Die00]. In turn, a spatial variation in ε_{xx} leads to position-dependent gain for (parasitic) TM-emission. [Yan95] even observed TM-gain surpassing the TE-gain in (unintentionally) stressed regions of ridge waveguide lasers of nominal TE emission.

Publications on high-power diode lasers mainly focus on polarization rotation when assessing the effect of strain on laser performance [Bie07, Cas13, Cas13a, Hol18]. For instance, [Cas13] extracted the spatial distribution of polarization rotation in mounted chips from photo-luminescence (PL) experiments, cf. Fig. 2.3(b,c). The data, of the therein defined rotated degree of polarization (ROP), reveal the anti-symmetric nature of the shear strain ε_{xy} resulting from processing and mounting. The effect was found on the two length scales for single-emitter chips ($\Delta x = 510\ \mu$m) and bar chips ($\Delta x = 9.9$ mm) in panels (b) and (c), respectively.

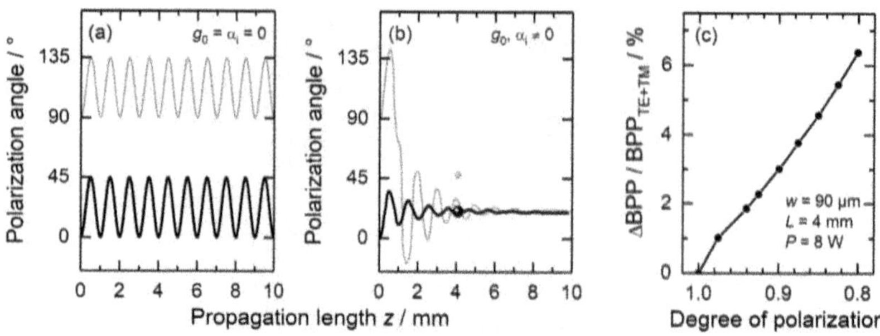

Figure 2.4: Effect of strain on emission properties. (a,b) Rotation of the axes of polarization upon propagation through a gain-/loss-less and an active BAL chip, respectively. The two cases of initially TE- (black) and TM-polarized (gray) emission are considered. [Hol18] (c) Effect of strain, quantified via the DOP, on the beam quality of a $w = 90\,\mu$m-wide BAL. [Win14]

[Hol18] gave a quantitative review of the rotation in a typical BAL in response to comparably large shear strain $\varepsilon_{xy} = 10^{-4}$, of which data is presented in Figure 2.4(a,b). Panel (a) treats the case of a gain- and loss-less chip ($g_0 = \alpha_\mathrm{i} = 0$) and lets initially purely TE- (black) or TM-polarized (gray) light propagate along a length z. The power is periodically interchanged between pure TE/TM polarization (0° and 90°, respectively) and TE-TM-mixed emission (45° and 135°, respectively). The situation changes for an active device, in which optical gain and losses are accounted for, cf. panel (b), with the dichroism $g_\mathrm{TE} = 2g_\mathrm{TM}$ assumed. The analysis further presumes that any TM-photon absorbed will relax into the heavy-hole band and be re-emitted as TE-photon. The periodic interchanging is still present, but both initially purely TE- (black) and TM-polarized (gray) emission are asymptotically rotated to the same polarization angle $\in (0 : 45)°$. That is, pure TM emission is partially rotated into the TE plane and partially absorbed and re-emitted as TE. The graph further includes two data points that mark the respective final polarization angle[7], after having traveled a length $L/(1 - R_\mathrm{f})$. From this, the angles of polarization rotation (initial to final) maximally to be expected in the devices studied in Section 5.1 are 20° and 43° for TE and TM emission, respectively.

Of the many publications devoted to processing- and mounting-induced strain in high-power diode lasers, few relate to the beam quality of the light emitted. In a study of $w = 90\,\mu$m-wide single-emitter lasers [Win14], for instance, the strain is quantified via the

[7]For this, photons in an $L = 4$ mm-long device with front reflectivity $R_\mathrm{f} = 2\%$ are stochastically emitted anywhere along the cavity and then on average travel a distance $L/(1 - R_\mathrm{f}) = 4.1$ mm. Thus the curves in panel (b) where averaged along this propagation length for the computation of the two data points.

overall DOP when biased and correlated with the BPP_{TE+TM} of the emission. Following their reasoning, strain worsens the beam quality as it adds TM emission, which itself is of low beam quality BPP_{TM} with contributions close to the emitter edges, and thereby compromises the overall beam quality by ΔBPP. Figure 2.4(c) displays the inferred relation and suggests that strain-induced beam-quality worsening remains below 10% for any DOP $> 80\%$.

The literature does not, however, provide data on the relation between strain and beam quality in high-power laser bars.

2.4. Thermal aspects of diode-laser operation

The thermal resistance R_{th} reflects the temperature rise of any device when operated and thereby dissipating the power P_{diss}.

$$R_{\mathrm{th}} = \frac{\mathrm{d}T}{\mathrm{d}P_{\mathrm{diss}}}$$

A theoretical prediction of its magnitude requires the thickness d and the thermal conductivity λ_{th} of the layer that connects the device to the (cooling) temperature reservoir, as well as its cross-sectional area $A = \Delta x \cdot \Delta z$. Under the assumption of one-dimensional heat flow (if $d \ll \min\{\Delta x, \Delta z\}$) the analytical expression

$$R_{\mathrm{th}} = \frac{d}{\lambda_{\mathrm{th}} A}$$

can be used [Col95]. Expanding this to a sequence of layers k (in series), with each the thickness d_k and thermal conductivity $\lambda_{\mathrm{th},k}$, as well as to laser bars with the lateral active dimension $\Delta x = n_{\mathrm{e}} w$ (n_{e} emitters of width w in parallel) and cavity length $\Delta z = L$, the formula reads

$$R_{\mathrm{th}} = \frac{1}{L\, n_{\mathrm{e}}\, w} \cdot \sum_k \frac{d_k}{\lambda_{\mathrm{th},k}}. \tag{2.5}$$

The following paragraph will study the deviation from one-directional heat flow as it allows lateral heat conduction as a minor perturbation to the major vertical heat path. The goal is to estimate the characteristic thermal length Δx_{th} across which the heating ΔT_{AZ} within an emitter decays towards its sides and reaches the reservoir temperature. The two lateral temperature tails of two neighboring emitters spaced at $d_{\mathrm{th}} = 2\Delta x_{\mathrm{th}}$ apart then do not overlap. This quantity will be necessary when studying the effect of thermal emitter cross-talk on bar beam-quality, cf. Sec. 5.2.

The geometry for this derivation is sketched in Figure 2.5 and assumes a single emitter

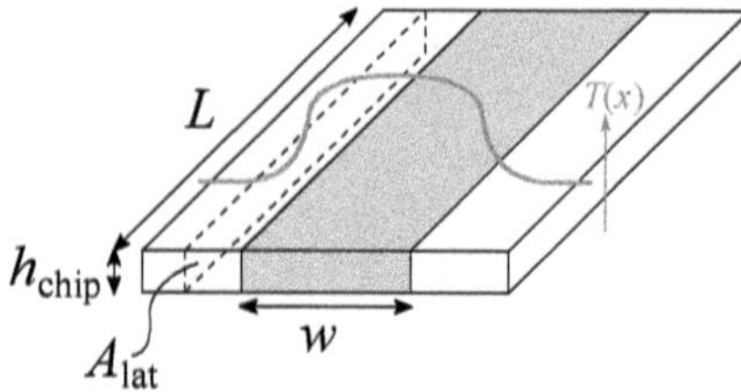

Figure 2.5: Sketch of an emitter stripe in a diode-laser chip with the lengths and the plane referred to during the derivation of the lateral thermal length d_{th}.

of width w processed onto an infinitely-wide and L-long semiconductor chip of the height h_{chip}. The relation between a heat-flux density distribution $\dot{\vec{q}}(\vec{r})$ and the corresponding temperature field $T(\vec{r})$ is given by Fourier's law [Fou22].

$$\dot{\vec{q}}(\vec{r}) = -\lambda_{th}(\vec{r}) \cdot \vec{\nabla} T(\vec{r})$$

Integrating the left part of the relation over the area A_{lat} enclosing one lateral side of the emitter delivers

$$\iint_{A_{lat}} \dot{\vec{q}} \cdot \mathrm{d}\vec{A} = \frac{P_{diss}}{n_e} \cdot \frac{h_{chip}}{2h_{chip} + w},$$

with P_{diss}/n_e the dissipated power of this one emitter. In the above, the following approximation was made: the dissipated power that flows through this one edge area A_{lat} relates to the total dissipated power as this edge area $h_{chip}L$ relates to the total heat conduction area $(2h_{chip} + w)L$. Integration of the right part of Fourier's law while assuming constant $\lambda_{th}(\vec{r}) = \lambda_{th,l}$ at the lateral boundary and defining the characteristic decay length Δx_{th} as the lateral distance across which the temperature drops by ΔT_{AZ} (average gradient) gives

$$-\iint_{A_{lat}} \lambda_{th} \vec{\nabla} T \cdot \mathrm{d}\vec{A} =: \lambda_{th,l} \frac{|\Delta T_{AZ}|}{\Delta x_{th}} \cdot h_{chip} L.$$

By using $P_{diss} = \Delta T_{AZ}/R_{th}$ and by taking into account that neighboring emitters need to be separated by $d_{th} = 2\Delta x_{th}$ in order to not interfere with each other (tails on each side) one finds

$$d_{th} = 2\lambda_{th,l} L n_e R_{th} (2h_{chip} + w).$$

Further, the sum in Eq. 2.5 is replaced by an effective value

$$L n_e R_{th} = \frac{1}{w} \left(\frac{d}{\lambda_{th}} \right)_{eff}$$

that covers realistic conditions in the studied bar-chip designs. Its value is derived from the thermal resistance R_{th} found in Section 4.2 and assumes[8] $(d/\lambda_{th})_{eff} = 2.1 \cdot 10^{-6}\,\mathrm{m^2K/W}$. Choosing a stripe width $h_{chip}/w \ll 1$ to justify the assumption that lateral heat conduction has only a minor effect on $(d/\lambda_{th})_{eff}$, e.g. $w = 1100\,\mu\mathrm{m}$ and using the values $\lambda_{th,l} \approx \lambda_{th,GaAs} = 47\,\mathrm{W/m\,K}$ [Sze07], $h_{chip} = 120\,\mu\mathrm{m}$ one finally arrives at

$$d_{th} = 2\,\lambda_{th,l} \cdot (1 + (2h_{chip})/w) \cdot (d/\lambda_{th})_{eff} \approx 240\,\mu\mathrm{m}. \tag{2.6}$$

If now two bar emitters are processed with a distance d_s in between each other one can distinguish the two cases

- thermally individual emitters if $d_s \geq d_{th}$,
- overlapping temperature profiles if $d_s < d_{th}$.

[8] $R_{th} = 0.05\,\mathrm{K/W}$, $n_e = 37$, $w = 186\,\mu\mathrm{m}$, $L = 6\,\mathrm{mm}$

3. Experimental Methods

This chapter introduces the experimental methods underlying the analysis of this work. Firstly, the fabrication of the devices under study, diode-laser bars, is outlined such that the reader gains a coarse overview of the most important steps. The second part is devoted to the used characterization techniques. The respective basic operation principle is followed by the possible parameter ranges and errors. Most importantly, Section 3.2.3 describes a test station for emitter-resolved beam-quality analysis of high-fill-factor bars. This novel test capacity was designed and constructed in the course of this work. Its description encompasses optical layout, mechanical stability, and long-term calibration stability.

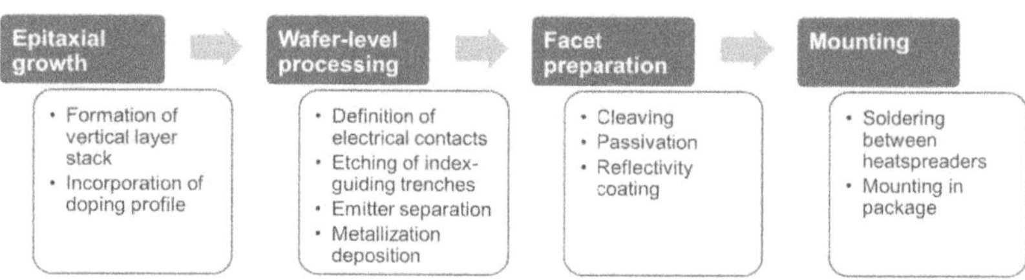

Figure 3.1: Illustration of the consecutive steps for the fabrication of high-power diode-laser bars.

3.1. Fabrication of broad-area laser bars

The consecutive steps for the fabrication of high-power diode-laser bars are illustrated in Figure 3.1 and can be coarsely divided into epitaxial growth, wafer-level processing, facet preparation and mounting. Photographs of the devices at selected process stages are shown in Figure 3.2. While [Erb00] gave a comprehensive review, only a few steps most critical for this work are described in the following.

The epitaxial (i.e. vertical) structure was grown via metal-organic vapor-phase epitaxy on (001)-oriented n-doped GaAs substrates [Bug16]. It comprises of a 3.5 μm-thick asymmetrically cladded AlGaAs waveguide core embedding an InGaAs single quantum well with a GaAsP barrier layer on each side. The structure is topped with a highly p-doped contact layer. The resultant vertical refractive index distribution (cf. Fig. 4.1) gives an overview of the grown layer stack.

Further, wafer-level processing includes layouting the bar emitters by definition of the electrical contacts via spatially selective implantation of the p-side contact layer. For this

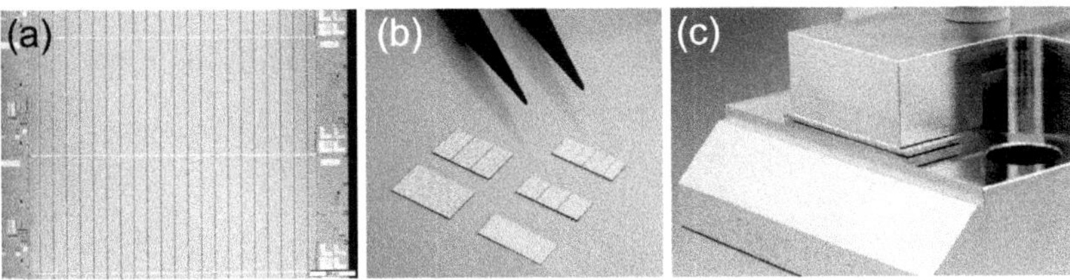

Figure 3.2: Photographs at different stages of laser-bar fabrication (a) Bars after completion of wafer-level processing. Selected areas of the wafer were cleaved off while the individual bars are still connected. (b) Fully processed bars after facet coating. (c) Conduction-cooled package with CuW heat-spreaders and the bar chip being mounted between the p-base and the massive n-contact. (©FBH/schurian.com)

"shallow" implantation, $^{4}He^{+}$ ions accelerated to a few tens of keV were directed onto the p-side wherein the electrical conductivity is significantly reduced upon penetration [Pea90], locally impeding current injection from above. As optional further process steps, either (i) another implantation process ("deep implantation") was utilized or (ii) lateral index-guiding trenches (IGT) were etched. In the first case, the implantation now reaches deeper into the vertical structure and was carried out outside the contact areas with an offset of 5 μm to the emitter edge with the use of $^{1}H^{+}$ and $^{4}He^{+}$ now accelerated to a few hundreds of keV. A simulation of the implantation effect onto the semiconductor material was carried out using [Zie17] and can be inspected in Figure 3.3(a). Parameters were chosen to reach an implantation depth of 200 nm above the active zone (AZ). As for option (ii), index-guiding trenches of rectangular shape were dry-etched with an offset of 5 μm to the emitter edges and to a depth similar to the implantation depth in (i) acting as $\Delta n_{\text{eff}} = 10^{-3}$ [Cru12]. The etched surfaces were then passivated using silicon nitride. Emitters with a width $w \geq 400\,\mu$m were additionally sub-structured with a periodic implantation pattern of longitudinal stripes into the p-contact layer ("shallow implantation") [Spr10, Kar17a, Cru17], cf. Fig. 3.3(b). All emitters were separated from each other with a separation trench etched down to the n-doped region. Ohmic contacts were then formed on the p-side and topped with a 3 μm-thick electro-plated Au layer. The wafer was then thinned down on its n-side to $\approx 120\,\mu$m thickness and equipped with a metallic contact. The so processed wafer was separated into individual bars by cleaving along the principal cubic symmetry planes. The chips were of 9.9 mm width and $L \in \{4, 6\}$ mm resonator length. The open facets were passivated with ZnSn at ultra-high vacuum conditions [Res05] and then anti- and high-reflection coated at the front and back side, respectively. Reflectivity values $R_{\text{f}} \in [1:2]\%$ and $R_{\text{r}} = 98\%$ were chosen. A topic of particular concern

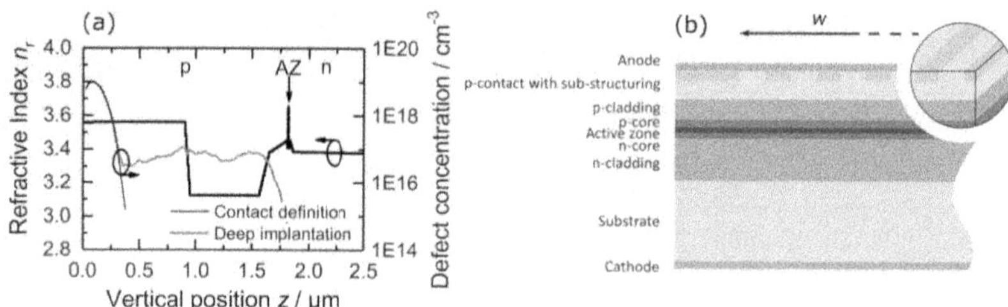

Figure 3.3: Chip processing details (a) Simulated implantation profiles along with the refractive index profile of the epitaxially grown layer stack. For contact definition and emitter sub-structuring, the shallow implantation is used. Deep implantation is an optional process step. (b) Illustration of the sub-structuring geometry applied for emitter widths $w \geq 400\ \mu\text{m}$. Depicted is a section of one sub-structured emitter.

for laser bars with a high-fill factor, as used in this work, is the absence of any defects or contamination on the laser facets, which would otherwise pose the risk of emitter failure (COMD, catastrophic optical mirror damage). This requires strict process control by inspection at critical steps and device selection based on a pre-defined catalog as was studied and proposed in [Ber18, Sch20].

In a last step, the bar chips were hard-soldered between thermal expansion-matched CuW heat-spreaders via AuSn solder [Wes08]. Using In solder, this sandwich was then mounted p-side down into a conduction-cooled package (CCP) with a massive n-contact all made of Au-plated Cu.

All fabrication steps were carried out at the FBH by experts in the respective departments according to the instructed design goals.

3.2. Characterization techniques

The following presents the different test methods used for the characterization of high-power diode-laser bars. The first setup described qualifies the grown and processed semiconductor material and outputs basic internal semiconductor laser parameters, such as modal gain g_{m} and internal loss α_{i}. Its basis is the commonly used length-dependent laser analysis [Col95]. Next, a test station capable of recording the output power–current–voltage dependence upon operation is detailed. It is required for demonstrating the power levels targeted and computing the conversion efficiency at which the light is emitted. The setup operates solely in QCW mode with variable duty cycle [0.1:4]% and feeds current up to 2 kA. A test station particularly crucial to this thesis is described thereafter. It is capable of measuring the beam quality of high-fill-factor bars on individual-emitter level and has been conceived in the course of this work. It allows to acquire near- and far-field profiles (both polarization-state-resolved) as well as wavelength spectra of the individual emitters at various operation points (variable current and duty cycle). Its optical and mechanical design had been guided by optical simulations [Epp18] allowing to quantify the component and assembly tolerances as well as the measurement fidelity. The last setup assesses the physical deformation of bar chips – also referred to as bar smile – resultant from internal and external strain sources.

3.2.1. Determination of internal laser parameters

Basic laser parameters, such as internal loss α_i and modal gain Γg_0, are inferred from diodes processed with different cavity lengths $L \in [1:6]$ mm and stripe widths $w = \{100, 200\}\,\mu$m in a short-loop process. Diode lasers processed in such a way do not exhibit full durability (limited heat removal, limited facet load), as the metallization-layer stack is simplified for faster processing and facets are neither passivated nor coated for anti-/high-reflection, cf. full-cycle processing described in Section 3.1. The chips are contacted with probes on the p-side and driven to $I = 2$ A with the output power P being recorded. Pulses with lengths $\Delta t_\mathrm{p} = 1\,\mu$s at a repetition rate $f_\mathrm{rep} = 5$ kHz are chosen hereby for low heating [Has14]. What follows is the cavity length-dependent laser analysis, for instance described in [Col95, Erb00]. Slope fitting of the output curves reveals the differential efficiency $\eta_\mathrm{d} = 2 \cdot (\mathrm{d}P/\mathrm{d}I) \cdot (e\lambda/hc)$ as a function of cavity length, cf. Eq. 2.2. From this, the internal loss α_i and the internal efficiency η_i are determined via fitting of

$$\frac{1}{\eta_\mathrm{d}}(L) = \frac{1}{\eta_\mathrm{i}} - \frac{\alpha_\mathrm{i}}{\eta_\mathrm{i}\ln(R)} \cdot L \quad \rightarrow \quad \alpha_\mathrm{i}, \eta_\mathrm{i}.$$

In another step, the threshold-current density $j_\mathrm{th} = I_\mathrm{th}/(Lw)$ inferred from the output curves is plotted as a function of the inverse cavity length to extract the modal gain Γg_0 from its slope

$$\ln(j_\mathrm{th}) = \ln(j_\mathrm{tr}) + \frac{\alpha_\mathrm{i}}{\Gamma g_0} - \frac{\ln(R)}{\Gamma g_0} \cdot L^{-1} \quad \rightarrow \quad \Gamma g_0, j_\mathrm{tr},$$

with j_tr the transparency current density, cf. Eq. 2.1. Comparing these inferred parameters to those from previous processing and growth runs allows for determining if the diode-laser material exhibits the desired properties and can thus be used for further full-cycle processing. Further, knowledge of those parameters lets one extrapolate the expected optoelectronic laser characteristic.

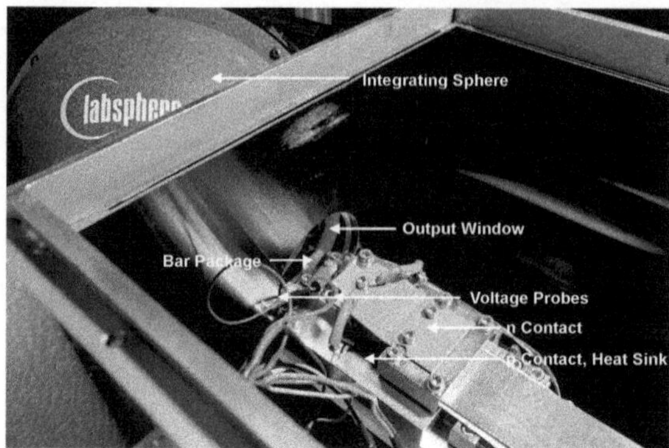

Figure 3.4: View onto the chamber holding the laser-bar package: The operation current is fed via the current rails on the right, while the laser emission exits the chamber to the left through the anti-reflection-coated window. The integrating sphere left to the chamber has a thermoelectrical detector, a photodiode with sufficient bandwidth (10 MHz), and a fiber port attached to it.

3.2.2. Electro-optical characterization and spectra

The basic electro-optical characterization of high-power bars comprises the measurement of operation current I, the emitted output power P, and the voltage V dropping across the device. This allows the electro-optical conversion efficiency η at the operation point I to be computed as

$$\eta(I) = \frac{P(I)}{I \cdot V(I)}.$$

The test station used enables electro-optical characterization in pulsed operation (pulse length $\Delta t_\text{p} \in [0.1 : 4]$ ms, duty cycle DC $\leq 4\%$) with currents up to 2 kA and simultaneously measures the wavelength spectra of the laser bar emission. A comprehensive description of its operation principle can be found in [Fre16, Fre19].

Figure 3.4 depicts the heart of the test station, with the measurement chamber holding the laser-bar package and an integration sphere. The operation current is fed through current rails for low electrical inductance in roughly rectangular pulse shape. The rails are equipped with an in-line resistor ($R_\text{in-line} = 205\,\mu\Omega$) for accurate current recording $I = V_\text{in-line}/R_\text{in-line}$. The voltage dropping across the device is measured with an extra pair of probes in a four-terminal configuration. The bar package used does not, however, allow contacting the chip itself but adds a package resistance $R \approx 10\,\mu\Omega$ to the measured voltage [Cru14]. The bar package is clamped onto the heatsink with a defined torque 40 cNm for reproducible thermal contact. The heatsink itself is flushed with silicone oil provided by a thermostat which is controlled via a temperature sensor (Pt1000) for constant heatsink temperature T_{HS}. The sensor is placed inside the heatsink underneath the clamped bar package. This configuration is capable of removing the maximum anticipated waste heat 40 W (time average). The emitted light passes an anti-reflection-coated window and is accepted by an integrating sphere. Its three ports hold a thermoelectrical detector measuring the time-averaged power, a photodiode with μs-range time response, and a fiber port for connecting a wavelength spectrometer. The thermoelectrical detector exhibits a comparably low wavelength sensitivity and is calibrated against absolute power standards (NIST, cf. [Cru13, Cru14]) in the 9xx nm-wavelength range. At the heart of the test station's operation principle is the time-dependent recording of current, voltage, and power with the in-line resistor, the voltage probes, and the photodiode being connected to an oscilloscope. While the current and voltage readings can be read off directly, the power signal from the photodiode $V_{\mathrm{PD}}(t)$ requires cross-referencing with the power P_{avg} measured by the thermoelectrical detector. A time trace of the power in absolute units $P(t)$ can be derived from $V_{\mathrm{PD}}(t)$, with a factor $\alpha(I)$ into which also the pulse-repetition rate $f_{\mathrm{rep}} = \mathrm{DC}/\Delta t_{\mathrm{p}}$ factors in. This factor needs to be computed separately at every operation point.

$$P(I,t) = \alpha(I) \cdot V_{\mathrm{PD}}(I,t) \text{ , with}$$

$$\alpha(I) = P_{\mathrm{avg}}(I) \;/\; \left(f_{\mathrm{rep}} \int_{\Delta t_{\mathrm{p}}} V_{\mathrm{PD}}(I,t')\,\mathrm{d}t' \right)$$

Values of power, current, and voltage given herein are now deduced from the plateaus of these pulse shapes in absolute units in the range of 65 to 95% pulse length ("gating") in order to ensure that the device has reached steady state (cf. quasi-CW operation).

$$P(I) = \frac{1}{0.3\Delta t_{\mathrm{p}}} \int_{0.65\Delta t_{\mathrm{p}}}^{0.95\Delta t_{\mathrm{p}}} P(I,t)\,\mathrm{d}t$$

The statistical error in power and efficiency is given as $\sigma_P/P \approx \sigma_\eta/\eta \approx 1\%$.
Further, the integrating sphere's fiber port is connected to a wavelength spectrometer for acquiring the spectrum of the laser emission at every operation point.

3.2.3. Test station for the analysis of bar beam-quality

This section describes a test station capable of analyzing the beam quality of high-fill factor laser bars, which has been conceived and realized in the course of this thesis [Kar20]. The targeted measurement quantities and required specifications are first stated, and then

their implications on the setup design are described. Calibration and comparison to reference is then outlined and finally the long-time stability under use is investigated.
Studies on the beam quality of high-fill-factor bars have been carried out by several groups [An15, Haa17, Str18]. In these investigations the bar emission has been characterized as a whole, indivisible beam. This is sufficient in the case of testing conformance with a particular bar specification. In contrast, in order to understand the beam-quality limitations specific to bars, it is necessary to probe the emission properties of every individual emitter processed onto a bar chip. With this method it is possible to assess a wide range of questions, including for example if and how neighboring emitters interact with each other and if the macroscopic dimensions of bar chips make them vulnerable to mechanical strain profiles. Hence, the projected test station shall be capable of measuring near-field profiles, far-field profiles, and wavelength spectra of every individual bar emitter, even when all emitters are biased simultaneously, as is typical during bar operation. The only previous concept realized divides the emission right at the output facet by blocking the beam emitted by all emitters but one via, e.g. glass prisms or knife edges [An14]. This, however, leads to unsatisfying separation in the case of small emitter distances ($d_s < 200\,\mu$m)[9], the latter being mandatory in high-power, i.e. high-fill-factor, bars. The herein proposed concept first magnifies the front-facet emission-pattern using a Keplerian telescope and images it onto a plane in which the, now magnified, near field profile can be separated properly using knife edges. The residual emission is then imaged through a converging lens placed at a distance of its focal length to the separation plane, thus generating the emission's angular distribution histogram in infinite distance, i.e. its far-field profile. Table 3.1 lists the required error ranges of the quantities to be measured and the anticipated operation conditions. This section will only focus on the requirements on the optical path, not on the current sources etc. required for bar operation at the listed conditions. The mechanical design of the test station shall further grant reproducible and easy-to-use changing between near-field and far-field configuration.

The near-field configuration

In the near-field configuration, the intensity distribution right at the facet of the BAL needs to be imaged onto a CCD in a magnified manner. A suitable arrangement is the Keplerian telescope, essentially comprising of only two lenses, cf. Fig. 3.5(a), of the focal lengths f_1 and f_4, respectively. Its basic operation principle can be derived using the matrix formalism, attributing particular transformation matrices to basic optical elements

[9]In a coarse estimate, the glass prisms need to be closer than $500\,\mu$m to the output facet for neighboring beams to not overlap at this emitter distance, assuming 10° divergence. This sets an unrealistic lower bound for the emitter distance and puts the front facet at risk.

Quantity	**Specification**
Near-field width (95% power content)	Error $\leq$ 2 μm
Far-field width (95% power content)	Error $\leq$ 0.1°
Polarization-axes resolution	
Wavelength resolution	
Emitter width w, as processed	[90:1200] μm
Expected full beam divergence Θ	Lateral $\leq 20°$
	Vertical $\leq 50°$
Emission wavelength λ	$\approx$ 940 nm
Bar optical power P	[0:1] kW
	([0:40] W time average)
Operation mode	Quasi-continuous wave
Pulse length Δt_{p}	[0.2:2] ms
Pulse repetition rate f_{rep}	[10:20] Hz

Table 3.1: Summary of the specifications expected from the designed test station in terms of measurement accuracy and anticipated properties of the incoming beam.

[Hal64]. The full derivation of the results to follow can be found in Appendix A. The single telescope's matrix reads

$$\mathrm{NF}_{\text{single}} = \begin{pmatrix} -\dfrac{f_4}{f_1} & 0 \\ \dfrac{z_1}{f_1 f_4} & -\dfrac{f_1}{f_4} \end{pmatrix}.$$

In the picture of light beams, a ray leaving the laser plane DL at position x_1 at an angle α_1, i.e. $\vec{x}_1 = (x_1\ \alpha_1)$, is imaged onto the CCD at position and angle

$$\vec{x}_2 = \mathrm{NF}_{\text{single}} \cdot \vec{x}_1 = \begin{pmatrix} -x_1 \dfrac{f_4}{f_1} \\ -\alpha_1 \dfrac{f_1}{f_4} + \dfrac{x_1 z_1'}{f_1 f_4} \end{pmatrix}, \tag{3.1}$$

with $z_1' = 4f_2 + z_1 + z_2$ in the sketched geometry, cf. Figs. 3.5(a,b). From the above, the magnification M of the telescope is read off as

$$M := \frac{x_2}{x_1} = -\frac{f_4}{f_1}, \tag{3.2}$$

with the negative sign indicating flipping of the image. Thus the magnification is given as the ratio of the lenses' focal lengths, while not depending on the detuning length z_1'. The latter does, however, affect the angle of incidence onto the imaging plane critically. An early version of the setup had been realized as such a simple Keplerian telescope with the optical components and the detuning length listed in Table 3.2. This setup did, however, not fulfill the required specification as near-field imaging worked only in a range of less than 400 μm, which does not suffice to image broad-area emitters with stripe widths of $w = 1200\ \mu$m, cf. Tab. 3.1. The shortcoming is illustrated in Figure 3.6(a), in which a simulation[10] of the beam paths is shown along with a measured near-field profile. The imaged diode-laser bar with 400 μm-wide emitters reveals a smooth intensity decay away from the optical axis. This effect is commonly attributed to vignetting [Hec09] in which any aperture of the setup, e.g. the limited (usable) diameter of lenses, is too small. As seen from the simulation, lens L4 with the clear aperture $\approx$ 41 mm may actually restrict outer beams. This effect does not, however, turn out to be the governing one, while actually the high angles at which the light strikes the CCD worsen the image. This effect was studied more closely such that the sensitivity curve as dependent on the angle of incidence was recorded. For this, a collimated beam parallel to the optical axis was scanned across lens L_4 and its intensity, as measured by the CCD, recorded. The setup and the as-

[10] This and following simulations were carried out with the application Beam Xpert DESIGNER [Epp18]. It offers selecting between propagating Gaussian beams acting in paraxial approximation and propagating Gauss-Schell-bundles for identifying aberrations.

(a) **Near-field configuration A**

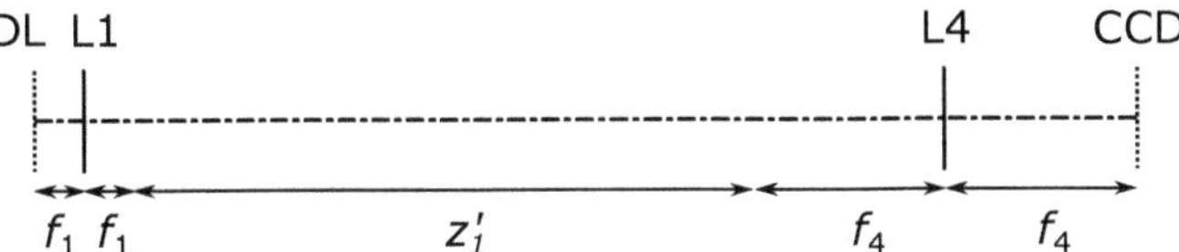

(b) **Near-field configuration B**

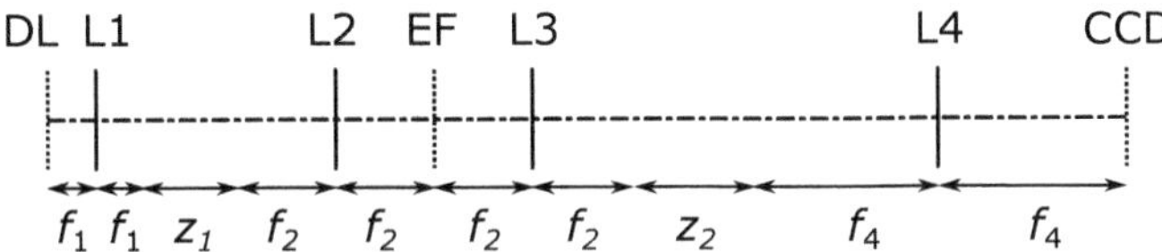

(c) **Far-field configuration**

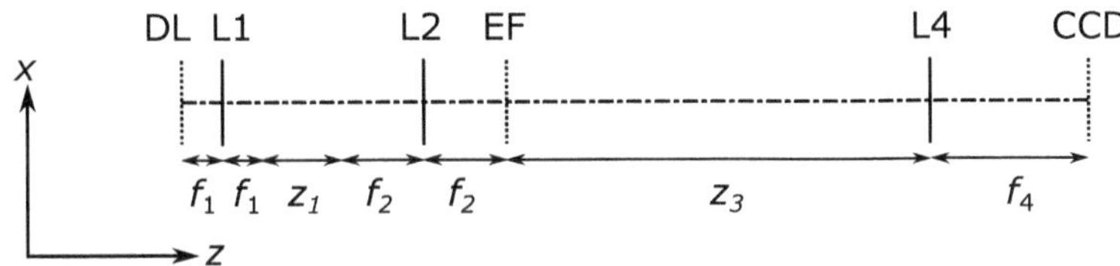

Figure 3.5: Coarse overview of the different lens configurations discussed when designing the beam-quality test station. The emission plane is denoted as DL (diode laser), the converging lenses as L1, L2, L3, L4 with their focal lengths $f_1, f_2, f_2, f_4 > 0$, and the plane of the emitter filter as EF. Additional optical lengths are z_1, z_1', z_2, z_3. (a) This Keplerian telescope with two converging lenses is a simple solution for imaging the near-field profile onto an image plane, e.g. a CCD. (b) This configuration features two consecutive telescopes, with the lenses L2 and L3 of equal focal lengths. This setup (b) has the light arriving at the image plane CCD at lower angles than in (a). Panel (c) displays the far-field configuration, in which a single telescope (L1,L2) magnifies the near-field onto a plane EF, in which the emission of a single bar emitter can be filtered out with an emitter filter. Lens L4 then images the angular distribution histogram onto the CCD.

measured sensitivity curve are depicted in Figure 3.7(a-b), with the rotational symmetry being apparent. Scanning across the lens makes the beam arrive at the CCD at a varying angle, so the as-measured sensitivity curve can be converted into an angular sensitivity curve, cf. Fig. 3.7(c), using elementary geometry. The dependence reveals a plateau in the range [-100:100] mrad and sharp flanks for angles beyond. Shadowing effects between adjacent CCD pixels may be accountable for this dependence. Concluding, an improved near-field setup should be designed such that the image is projected onto the CCD within this angular range, even for wide emitters 1200 μm. This requirement is fulfilled by a configuration comprising of two consecutive telescopes, cf. Fig. 3.5(b). The two additional lenses inserted in between need to act only in the lateral direction (cylindrical lenses) and exhibit the same focal length. The imaging matrix of this setup reads (cf. App. A)

$$\mathrm{NF}_{\mathrm{double}} = \begin{pmatrix} \frac{f_4}{f_1} & 0 \\ -\frac{z_1+z_2}{f_1 f_4} & \frac{f_1}{f_4} \end{pmatrix}.$$

The magnification factor $M = f_4/f_1$ is still determined by the outer lenses' focal lengths as the inner lenses' equal focal lengths cancel out. The insertion of the second telescope does, however, affect the angle of incidence α_2 at which the incoming profile will be imaged.

$$\alpha_2 = -\left(-\alpha_1 \frac{f_1}{f_4} + \frac{x_1(z_1+z_2)}{f_1 f_4}\right)$$

Comparing to the previous two-lensed setup, cf. Eq. 3.1, the angle of incidence is now diminished by the amount $(4x_1 f_2)/(f_1 f_4)$. In absolute figures treating the limiting case of a beam originating from the edge of a 1200 μm-wide emitter ($x_1 = 0.6$ mm), its angle of incidence is reduced from ≈ 410 mrad to ≈ -40 mrad, when lenses of $f_2 = 150$ mm are used. Thus this signal, previously attenuated by the CCD, is now counted properly as apparent from the recorded sensitivity curve Fig. 3.7. Signals stemming from distances closer to the optical axis are imaged at even lower angles. These coarse considerations are verified by both the simulated beam paths and the resultant measurement of a near-field profile shown in Figure 3.6(b). The measurement shows that at least in a range $\pm 1000\,\mu$m off the optical axis the profile is imaged properly, demonstrating the suitability for the intended purpose ($w_{\mathrm{max}}/2 = 600\ \mu$m). Now the choices made are checked against the vertical boundary conditions. As in the required specifications in Table 3.1, the divergence of the lasers to be characterized measures up to 50°. The selected lens L_1 (Tab. 3.2, $\mathrm{NA} = 0.54 \rightarrow 2 \cdot \arcsin(0.54) = 65°$) is capable of collecting all the emitted light. Once collimated, the beam has a spatial extent of $2 \cdot f_1 \cdot \tan(50°) = 17\,$mm which puts a lower limit on the diameter and opening of lenses and apertures in the setup.

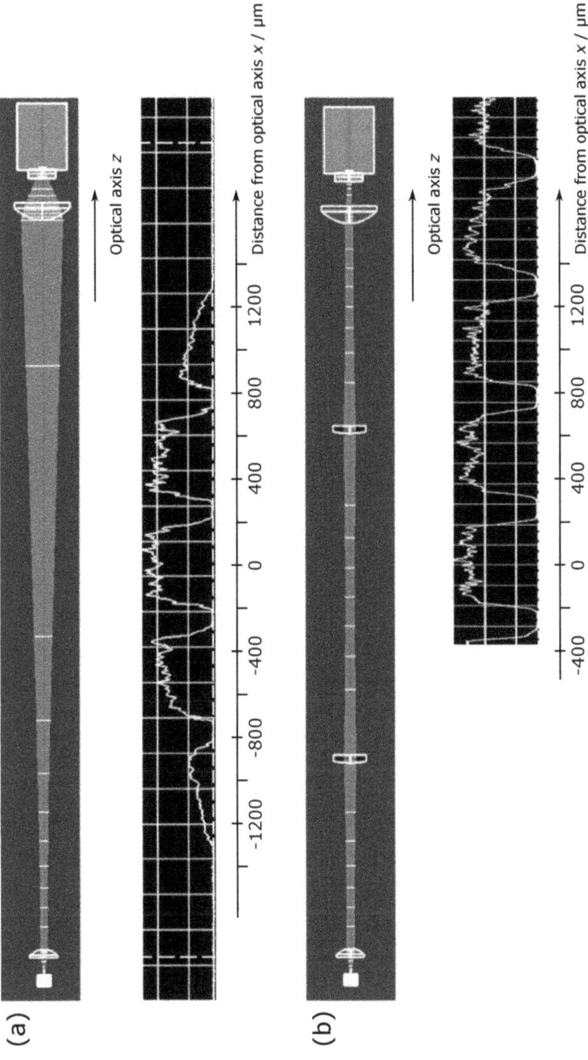

Figure 3.6: Optical simulations of two telescope configurations and the corresponding measured near-field image. (a) A simple Keplerian telescope makes the light arrive at the CCD at high angles of incidence and thus wide emitters are not imaged properly, as the used CCD camera exhibited angle-dependent sensitivity. (b) This is solved when using two consecutive telescopes with the inner (cylindrical) lenses of equal focal lengths.

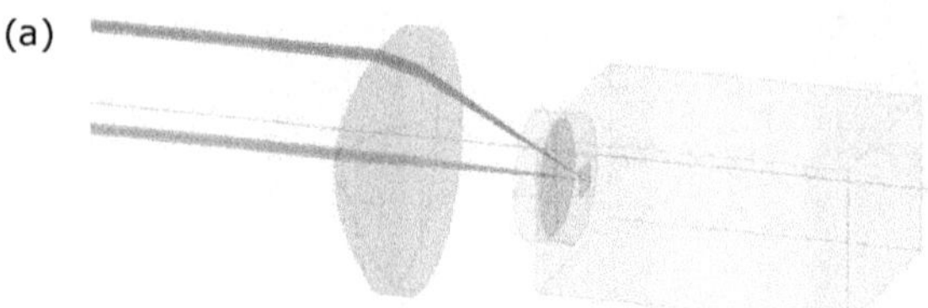

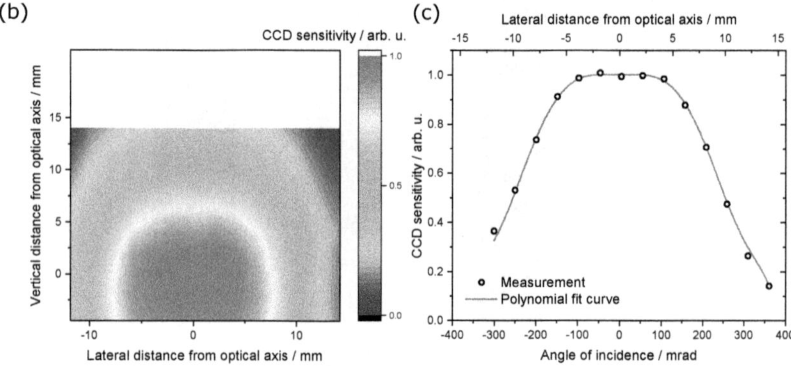

Figure 3.7: Determination of the CCD camera's angle-dependent sensitivity. (a) A collimated beam parallel to the optical axis is scanned across a lens and focused to the lens' focal point, thereby varying its angle of incidence. (b) The intensity as recorded by the CCD is measured as a function of position away from the optical axis. (c) This dependence can be converted to a function of the angle of incidence. The sensitivity curve exhibits a plateau in the angular range [-100:100] mrad and flanks for angles beyond.

Component	**Specification**	**Model**
Single telescope		
Lens L_1 Lens L_4	$f = 20$ mm, NA= 0.54, clear aperture ≥ 20 mm $f = 40$ mm, NA= 0.55, clear aperture ≥ 41 mm	Asphericon A25-20HPX A50-40HPX
CCD camera	Pixel size $\Delta x = \Delta y = 3.69\ \mu$m	PointGrey GS3-U3-28S4M-C
Detuning length	$z_1' \approx 545$ mm	
Additional in double telescope		
Lenses L_2, L_3	$f_{\text{lateral}} = 150$ mm, dimensions 30×32 mm×mm	Thorlabs LJ1629L1-B
Detuning lengths	$z_1 \approx 0$ mm, $z_2 \approx -55$ mm	

Table 3.2: Components with specifications as used in the two described near-field imaging systems comprising of either a single Keplerian telescope or a doubled telescope, cf. Fig. 3.5(a-b).

Lastly, the proposed setup design is checked against its resolution. The diffraction-limited resolution is mainly determined by the first light-collecting object, here L_1 and is given as [Ray79][11]

$$\Delta x_{\text{res}} = \frac{0.61\ \lambda}{\text{NA}} \approx 1\ \mu\text{m},$$

at $\lambda = 940$ nm. The finite pixel size of the used CCD camera defines another limit for the resolution. Divided by the factor of magnification $M = 2$, one obtains

$$\Delta x_{\text{res}} = 3.69\ \mu\text{m}\ /\ 2\ \approx\ 2\ \mu\text{m},$$

which is larger than the diffraction limit, but still in line with the desired specification, cf. Tab. 3.1.

The far-field configuration

The configuration for the acquisition of emitter-resolved far-field profiles is depicted in Figure 3.5(c). It uses the same optical assembly as for near-field profiles, except for lens L3 being removed. The Keplerian telescope L1-L2 images the near field onto the emitter-

[11] From the original angular definition, a spatial one is deduced and combined with the approximation NA $\approx \varnothing/2f$, with $\varnothing$ the lens' clear aperture.

filter plane EF with the magnification

$$M_{\mathrm{EF}} = f_2/f_1 = 7.5,$$

thus emitters of widths $w \in [90 : 1200]\,\mu\mathrm{m}$ measure about $[0.7 : 9]$ mm in the plane EF. This defines the opening of the emitter filter required when fully opened ($\approx 10\,\mathrm{mm}$), here realized as laterally movable knife edges. Blocking the emission of all emitters but one then allows generating an emitter-resolved far-field using lens L4. The imaging matrix is derived in the appendix (cf. Eq. A.2) and contains the relationship between far-field angle α_1 and position on the CCD x_2

$$|\alpha_1/x_2| = \frac{f_2}{f_1 f_4} = 0.188\,\mathrm{rad/mm} = 10.75\,°/\mathrm{mm}, \tag{3.3}$$

which is the essential calibration factor for this configuration.
In the next step, it is tested if the experimentally determined angular sensitivity curve of the CCD camera may affect the far-field image. As derived in Equation A.1, the angle of incidence onto the camera α_2 is given as

$$\alpha_2 = \frac{f_2}{f_1 f_4} \cdot x_1 - \alpha_1 \frac{f_1}{f_2}(1 - \frac{z_3}{f_4}).$$

Scanning through the parameter space $x_1 \in [-0.6 : 0.6]$ mm, $\alpha_1 \in [-20 : 20]°$ (taken from the required specifications), not all angles of incidence α_2 fall into the safe range $[-100 : 100]$ mrad that was deduced from Figure 3.7(c). The angular sensitivity may thus indeed affect the image. In a coarse estimate assisted by [Epp18], the sensitivity curve is separately applied to both a reference profile obtained from a $186\,\mu$m- and from a $1200\,\mu$m-stripe emitter. From the change in profile width one can estimate the significance of the sensitivity function in this far-field configuration. In the case of the narrower emitter, the profile does not change at all ($\Delta < 0.01°$ in 95%-width), whereas the wider emitter shows a profile that was narrowed by $\Delta \leq 0.2°$, however, in the improbable extreme case of a 20°-wide profile ($\rightarrow \Delta \leq 1\%$).
The angular resolution $\Delta\theta_{\mathrm{res}}$ afforded by the setup can be estimated from the CCD pixel width and Eq. 3.3 as $\Delta\theta_{\mathrm{res}} = 3.69\,\mu\mathrm{m} \cdot 10.745°/\mathrm{mm} = 0.04°$. Along with the influence of the CCD camera sensitivity, as just estimated, one can state the resolution

$$\Delta\theta_{\mathrm{res}} = 0.1°.$$

Mechanical setup design

The optical design as detailed so far is illustrated in Figure 3.8(a), taken from an optical

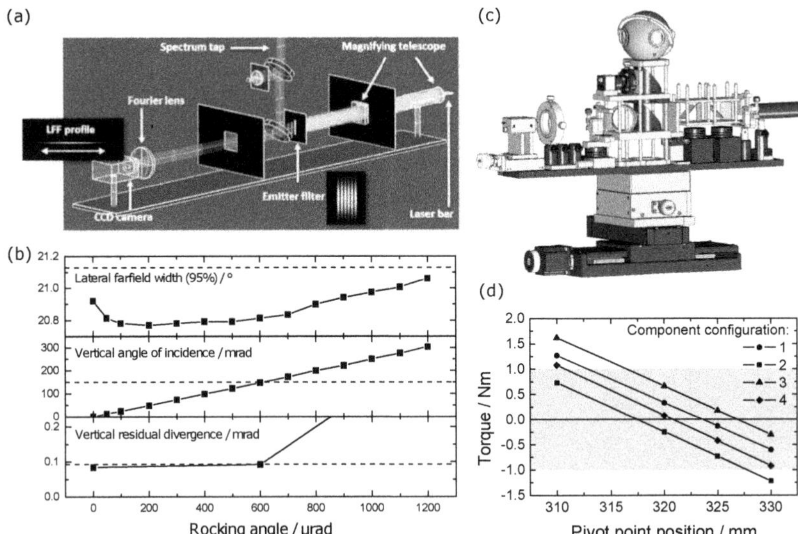

Figure 3.8: Analysis of the mechanical stability of the beam-quality setup. Displayed is (a) a simulation of the optical beam path and (b) the effect of rocking of the setup with respect to the incoming laser beam. Assembly (c) of purchasable and custom-made components to hold optics, along with the torque (d) acting on the translation stages.

simulation of the beam path using [Epp18]. It is now analyzed regarding its mechanical boundary conditions.

As the setup's intended use is the characterization of laser bars, the entire optical beam path is placed on a movable breadboard which will be translated from emitter to emitter. The beam path being movable bears the risk of the bread board rocking, which was observed to alter the recorded image significantly. This is why a stability analysis is carried out in the following via optical simulation. For this, rocking of the bread board is emulated by varying the angle between laser-emission forward-direction and the optical axis (nominally vanishing) as well as setting the corresponding height offset. A 1200 μm-stripe laser with a divergence $\approx 20°$ is propagated through the setup and three important optical quantities are recorded, being (i) the simulated lateral far-field width on the CCD chip, (ii) the average vertical angle of incidence onto the CCD camera (cf. angular sensitivity dependence), and (iii) the vertical residual divergence of the collimated beam right after the first lens L1. All three quantities are displayed in Figure 3.8(b) as a function of

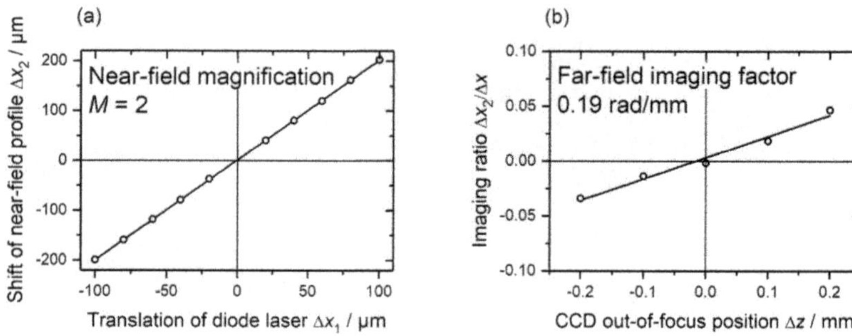

Figure 3.9: Measurement of the imaging factors in near- (a) and far-field (b) configuration. The experimentally determined values compare well to the ones deduced theoretically.

the rocking angle, along with the maximally tolerable deviation from nominal conditions (dashed lines). From this analysis, the mechanical design of the test station must guarantee to limit rocking below $\approx 600\,\mu$rad. A mechanical design comprising all the components to hold the optical design as simulated, along with three translation stages, is depicted in Figure 3.8(c). From this, one can estimate further requirements on the translation stages. These need to bear a mass of $\approx$ 10 kg (plus the weight of translation stage(s) above, if relevant) and a torque of $\approx$ 1 Nm. The latter was deduced from the masses and positions of all components. Figure 3.8(d) shows the acting torque as a function of pivot point at which the bread board is supported by the upmost translation stage. Further the torque is displayed for four different configurations the setup will be used at, e.g. with or without polarizer.

Determination of the imaging factors

In the following, the constructed measurement station is tested experimentally against its imaging factors. In case of the near-field configuration, this is the image magnification. For this, a $90\,\mu$m-stripe laser is translated laterally, with respect to the optical axis, and at every position its near-field profile is recorded. A shift Δx_1 in real space then corresponds to a shift $\Delta x_2 = M \cdot \Delta x_1$ of the near-field's center of mass as measured on the CCD. Figure 3.9(a) shows the acquired dependence from which the magnification is determined as its slope $M = 2.003 \pm 0.005$. This resembles the value expected from Equation 3.2, $M = 2$.

The imaging factor of the far-field configuration, i.e. the relation between far-field

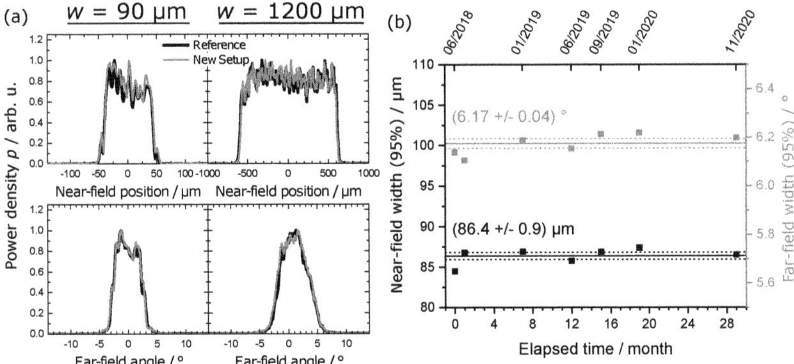

Figure 3.10: Determination of the measurement accuracy and precision. (a) Near- and far-field profiles are cross-calibrated at two stripe widths w. (b) The repeatability of the measurement is tested over the course of 29 months. Solid and dashed lines represent mean value and standard deviation, respectively.

angle and position on the CCD, is acquired via the procedure detailed in the appendix, culminating in Equation A.3. Again, the laser is translated laterally (Δx) and now the shift in far-field position (Δx_2) is recorded at different longitudinal positions z the CCD camera is placed at. The slope of this dependence corresponds to the imaging factor looked for. Figure 3.9(b) shows the result of this analysis, with the determined slope being $0.194 \pm 0.016\,\mathrm{rad/mm}$. The theoretically deduced value ($0.188\,\mathrm{rad/mm}$, cf. Eq. 3.3) lies in this error range.

The two imaging factors determined for the two setup configurations can further on be used in the evaluation of recorded near- and far-field profiles.

Comparison against reference and temporal stability

The near- and far-field profiles measured by the as-constructed test station are now compared against trusted references. For this, one single-emitter device of stripe width $w = 90\ \mu\mathrm{m}$ and one of $w = 1200\ \mu\mathrm{m}$ were tested for cross-calibration purposes. Profiles acquired with the new setup and with the reference are shown in Figure 3.10(a) for both stripe widths. The profiles are reproduced well, with the deviations in the deduced profile widths (cf. Tab. 3.3) in line with the required setup specifications. Near-field profiles were reproduced to within $2\ \mu\mathrm{m} \mathrel{\hat{=}} 2\%$ ($w = 90\ \mu\mathrm{m}$) and $14\ \mu\mathrm{m} \mathrel{\hat{=}} 1.2\%$ ($w = 1200\ \mu\mathrm{m}$) and far-field widths deviated by $0.1°$ in both cases.

Lastly, the temporal stability of the test station was recorded over the course of 29 months,

Stripe width w / μm	**90**	**1200**
Deviation in lateral near-field width (95%)	2 μm / 2%	14 μm / 1.2%
Deviation in lateral far-field width (95%)	0.1° / 1.6%	0.1° / 1.2%

Table 3.3: Measurement accuracy of the constructed beam-quality test station. Displayed is the deviation in 95%-width from reference measurements.

cf. Fig. 3.10(b). The standard deviation in near-field and far-field width was determined as 0.9 μm and 0.04°, respectively.

Deduction of bar far-field profiles

The overall bar far-field profile $p(\theta)$ is deduced from the individual emitter far-field profiles $p_i(\theta)$ via elementary summation, which resembles the incoherent superposition of n_e emitter's radiation.

$$p(\theta) = \sum_{i=1}^{n_e} p_i(\theta)$$

3.2.4. Measurement of bar smile

The concept of laser bars assumes ideally flat chips of semiconductor material. As-processed, these devices do, however, exhibit deviations from planarity. The spatial profile when looked at from the front facet is commonly referred to as the bar's smile profile and is used to describe the plastic deformation experienced by the chip. In order to measure such profiles, one needs to capture vertical positions within the sub-micron to few-micron range along the entire bar width of $\approx$ 10 mm. One way to capture these two different orders of magnitude is the use of different magnification factors for the vertical and the lateral direction, respectively, realized via the use of cylindrical lenses, each acting only in one direction.

In the used setup, the bar is biased well below its threshold ($\approx$ 3 A, CW) and the emission is imaged onto a screen for recording. Vertical imaging is done with a magnification of $M \approx 2000$ previously determined in calibration runs. Special care has to be devoted to ruling out that the used delicate cylindrical lens, with sub-mm focal length, exhibits a deformation profile itself. The recorded image essentially reveals the two-dimensional front-facet emission-pattern with the individual emitters being visible along with their

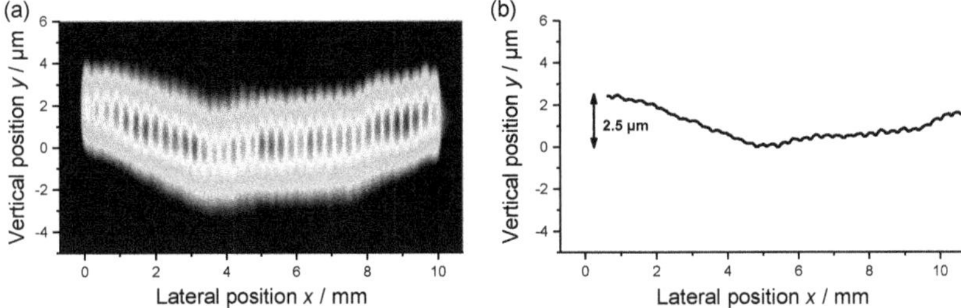

Figure 3.11: Illustration of how smile profiles are measured. (a) The front-facet emission-pattern is acquired with different magnification factors in vertical and lateral direction. (b) From this, the emitter location is deduced as a function of lateral position. The displayed smile profile spreads 2.5 μm.

bending by the chip deformation, cf. Fig. 3.11(a). When computing the vertical center of mass for each position, one can generate the vertical emitter position as a function of lateral position, i.e. the smile profile, as seen in Figure 3.11(b). The bar design used for this illustration features 37 emitters of 186 μm stripe width spaced at 250 μm emitter pitch. The ripples in the deduced profile are artifacts from the sharp intensity drop between adjacent emitters and do not stem from actual chip undulations. Now, a simple but common way to quantify smile is to measure the spread in the smile profile, measuring 2.5 μm in the shown example, to be trusted to ± 0.1 μm.

4. Diode-Laser Bars for Highly Efficient 1 kW-Emission

This chapter describes the development of highly efficient laser bars emitting a power of $P = 1\text{kW}$ at room temperature. Firstly, the EDAS-vertical-structure [Has14] is introduced as a concept enabling highly-efficient emission via low areal series resistance $R_s \cdot A$ and low optical loss α_i. Secondly, this chapter studies the temporal behavior of the laser emission on the millisecond-scale. The gained knowledge is applied to infer the efficiency at varying thermal conditions. Next, the longitudinal structure is varied via a change in cavity length, exploring the potential for efficiency increases. Lastly, the impact of the lateral design is studied, by simultaneously changing emitter width and fill-factor. In all development steps, the underlying idea is first introduced and then its potential numerically extrapolated, followed by measurement results from processed devices.

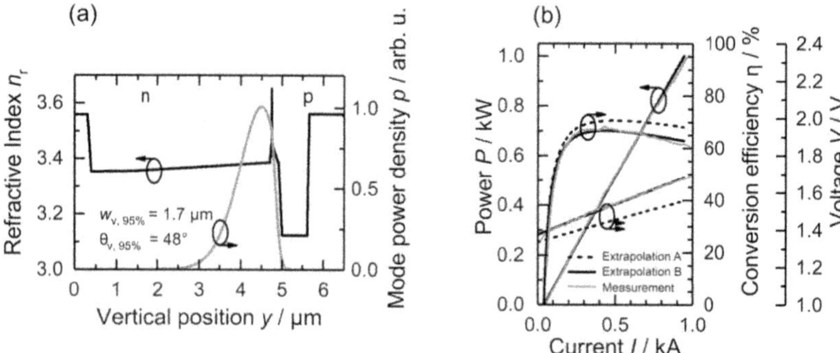

Figure 4.1: Vertical structure suitable for 1 kW-emitting laser bars. (a) Waveguide of a double-asymmetric design and the computed supported fundamental mode. (b) Power-voltage-current extrapolation for this vertical structure and measurement result of a processed laser bar.

4.1. Vertical structure for high-efficiency operation

This section introduces the vertical structure chosen in this work to realize highly-efficient 1 kW-emission. The design is based on the concept of so-called extreme double-asymmetric (EDAS) large optical cavities. Not attempting to repeat all details explained in [Has14], this vertical design features (i) highly asymmetric cladding layer compositions and (ii) a very thin waveguide p-side core. Both characteristics are also apparent from the plot of the refractive index as a function of vertical position in Figure 4.1(a). Using an FBH-developed

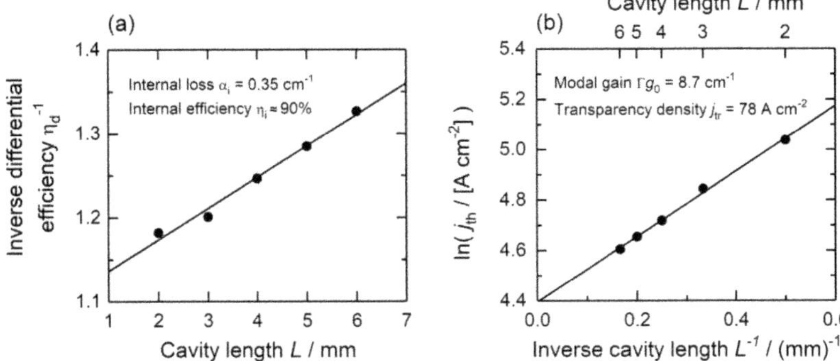

Figure 4.2: Cavity length-dependent analysis for the retrieval of laser-internal parameters. (a) Inverse differential efficiency η_d^{-1} as a function of cavity length L. (b) Natural logarithm of the threshold current density j_th as a function of inverse cavity length L^{-1}.

simulation tool [Wen90, Wen13] the fundamental mode supported by this waveguide was computed and is plotted alongside. The specific modal confinement provided leads to a vertical mode that (i) slowly decays towards the n-side, while abruptly vanishing towards the p-side and (ii) does not peak in the active zone but within the n-side. This design has distinct advantages over, previously developed, symmetric designs, such as reduced

- areal series resistance $\rho_\mathrm{a} = R_\mathrm{s} \cdot A$ due to the thin, inherently resistive, waveguide p-side core,
- internal loss α_i as result of low modal overlap with the p-side,
- power saturation $\mathrm{d}^2P/\mathrm{d}I^2 < 0$ via mitigated voltage-driven carrier leakage.

As simulated, the vertical mode reveals a near-field width of $w_\mathrm{v} = 1.7\,\mu\mathrm{m}$ and far-field width of $\theta_\mathrm{v} = 48°$ (both 95% power content).

Broad-area lasers of this vertical structure were grown and processed as outlined in Section 3.1 and then tested at different cavity lengths. This allows retrieving the internal laser parameters, cf. Sec. 3.2.1. Figure 4.2 depicts the results of this analysis, from which the internal loss $\alpha_\mathrm{i} = 0.35\,\mathrm{cm}^{-1}$, the internal efficiency $\eta_\mathrm{i} \approx 90\%$, the modal gain $\Gamma g_0 = 8.7\,\mathrm{cm}^{-1}$, and the transparency current density $j_\mathrm{tr} = 78$ A cm^{-2} were deduced via linear fits. This analysis does not, however, allow assessing the electrical behavior owing to the processing details of such short-loop process lasers.

These parameters can now be used to estimate the performance of this vertical structure

Parameter		Extrapol. A	Extrapol. B
Cavity length	L / mm	4	
Number of emitters	n_e	37	
Emitter width	w / μm	186	
Facet reflectivity	R_f, R_r / %	2, 98	
Voltage offset	V_0 / V	1.333	
Defect voltage	V_d / mV	0	40 [Fre19]
Series resistance	R_s / mΩ	0.25	0.33 [Fre19]
Areal series resistance	ρ_a / Ω (mm)^2	6.6	9.1
Modal gain	Γg_0 / cm^{-1}	8.7	
Internal loss	α_i / cm^{-1}	0.35	
Internal efficiency	η_i / %	90	
Transparency current density	j_{tr} / A cm^{-2}	78	
Thermal resistance	R_{th} / K W^{-1}	0.015 [Kar19]	
Threshold-characteristic temperature	T_0 / K	150 [Kau19]	
Slope-characteristic temperature	T_1 / K	600 [Kau19]	

Table 4.1: Parameters used for the optoelectronic performance extrapolation with results shown in Fig. 4.1(b).

as kW-class laser bars, with the result being presented in Figure 4.1(b). Table 4.1 lists all values assumed for this extrapolation. In particular, the electrical properties were estimated in two different ways, termed extrapolation A and B, respectively.

$$V(I) = V_0 + V_d + R_s \cdot I$$

In extrapolation A, the voltage at vanishing current $V(0)$ was simply estimated from the lasing wavelength $\lambda \approx 930$ nm, as the minimum flatband voltage

$$V_0 = \frac{hc}{e\lambda},$$

and the series resistance R_s calculated from the individual layer's material properties. Extrapolation B accounts for an extra defect voltage V_d, arising from non-ideal interfaces and electrical potential barriers between different-bandgap semiconductors [Fre19], as well as an increased series resistance due to incomplete carrier capture in the active zone. The details of this mechanism are beyond the scope of this work, but [Fre19] found an increase of about a third for an active-zone/waveguide-core interface of similar material composition as used herein, $0.25\,\text{m}\Omega \cdot 4/3 = 0.33\,\text{m}\Omega$. This corresponds to an areal series resistance of $\rho_a \approx 9\,\Omega\,(\text{mm})^2$.

Measurement results obtained from a processed laser bar in low-thermal QCW conditions (pulse length $\Delta t_p = 200\,\mu$s, repetition rate $f_{rep} = 10$ Hz) are plotted alongside these two

$I(1\,\mathrm{kW})$ / A	$V(1\,\mathrm{kW})$ / V	$\eta(1\,\mathrm{kW})$ / %
975	1.69	61

Table 4.2: Measurement result of the 4 mm-long, 69%-fill-factor bar whose power-voltage-current characteristic is depicted in Fig. 4.1(b).

extrapolations in Figure 4.1(b), showing a good match for the values of V_d and R_s chosen in extrapolation B. This 4 mm-long bar with 37 emitters of width $w = 186\,\mu\mathrm{m}$ emits 1 kW at an efficiency of 61%, cf. Tab. 4.2 [Str17]. This device serves as the baseline design for the following studies in this work.

4.2. Thermal transient

This section investigates the temporal behavior of the lasing emission when the bars are driven in QCW-mode. In particular, the millisecond-scale heating process is studied and then modeled in a rate equation-based approach.
For this, a kW-class bar of length $L = 6\mathrm{mm}$ is driven at varied duty cycle $\mathrm{DC} = \Delta t_\mathrm{p} \cdot f_\mathrm{rep}$, with the emitted power P, the fed current I, the voltage V dropping across the chip, and the emission wavelength λ being recorded, cf. Fig. 4.3. Here, the repetition rate $f_\mathrm{rep} = 10$ Hz was held constant, while the pulse width Δt_p was varied from 0.1 to 4 ms. As seen in panel (a) the optoelectronic characteristic hardly changes in this case, while the heating, cf. panel (b), does increase with pulse length. For this graph, the dissipated power

$$P_\mathrm{diss}(I) = I \cdot V(I) - P(I)$$

was calculated at every operation point and matched with the corresponding lasing wavelength λ. Now, a shift in wavelength $\Delta\lambda$ can be converted to an increase in active-zone temperature

$$\Delta T_\mathrm{AZ} = \Delta\lambda \cdot \left(\frac{\mathrm{d}\lambda}{\mathrm{d}T}\right)^{-1},$$

with the conversion factor $\mathrm{d}\lambda/\mathrm{d}T \approx 0.348$ nm/K, originating from the temperature-dependent band-gap energy [Var67] approximated for the used temperature range and experimentally verified for this wavelength, e.g. in [Win14]. The pulse length-dependent thermal resistance, i.e. thermal impedance Z_th, is deduced via linear fits

$$Z_\mathrm{th} = \frac{\mathrm{d}T_\mathrm{AZ}}{\mathrm{d}P_\mathrm{diss}},$$

and is plotted in Figure 4.4(a) as a function of the pulse length Δt_p. The non-linear behavior can be understood as the competing interplay of the laser's self-heating P_diss

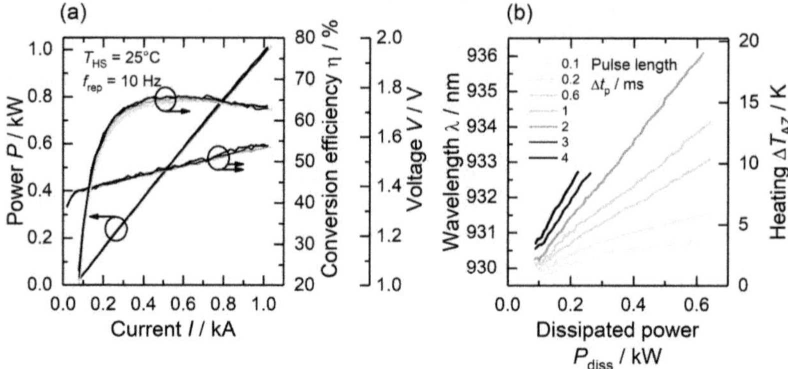

Figure 4.3: Duty cycle-dependent analysis of laser-bar performance. (a) Power P, Voltage V, conversion efficiency η as a function of current I in QCW-mode for pulse widths $\Delta t_\mathrm{p} \in [0.1 : 4]$ ms at constant repetition rate $f_\mathrm{rep} = 10$ Hz. (b) Central wavelength λ of the spectra measured alongside, now plotted as a function of dissipated power P_diss.

and the cooling across a thermal resistance R_th via thermal contact to a heatsink. The situation is sketched in Figure 4.4(b). A body of the thermal capacity C_th in J/K is subject to the thermal energy $P_\mathrm{diss}\mathrm{d}t$ upon which its temperature increases by $(P_\mathrm{diss}\mathrm{d}t)/C_\mathrm{th}$ above ambient temperature $T^{(0)}$. Thermal contact to an infinite reservoir with ambient temperature $T^{(0)}$, i.e. the heatsink, creates a gradient that results in a heat flux $-(T - T^{(0)})/R_\mathrm{th}$, reducing the body's temperature by $(T - T^{(0)})/(R_\mathrm{th}C_\mathrm{th})$. Cast in a differential equation, the situation reads

$$\frac{\mathrm{d}T(t)}{\mathrm{d}t} = +\frac{P_\mathrm{diss}}{C_\mathrm{th}} - \frac{T(t) - T^{(0)}}{R_\mathrm{th}C_\mathrm{th}}, \qquad T(t=0) = T^{(0)}.$$

From this, the temporal behavior of the body's heating $\Delta T = T - T^{(0)}$ is derived as

$$\Delta T(t) = P_\mathrm{diss} \cdot R_\mathrm{th} \cdot \left(1 - \exp\left(-\frac{t}{R_\mathrm{th}C_\mathrm{th}}\right)\right), \qquad \tau_\mathrm{th} = R_\mathrm{th}C_\mathrm{th},$$

with its characteristic time constant τ_th. For infinite times $t \to \infty$, this dependence convergences to the CW-value $P_\mathrm{diss} \cdot R_\mathrm{th}$ of the probed volume, regardless of its thermal capacity C_th.

This transient can now be assessed by varying the pulse length Δt_p and fitting to the acquired measurement data, cf. Fig. 4.4(a). Therein, the time-dependent thermal resistance

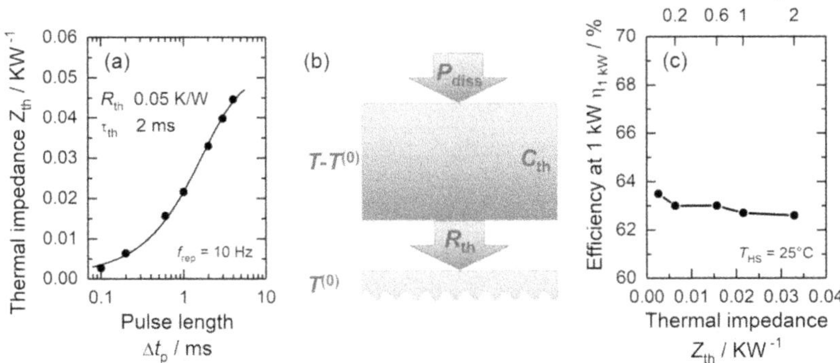

Figure 4.4: Thermal transients in kW-class laser bars, here $L = 6$ mm. (a) Measured data points and fitted curve of the thermal impedance Z_{th} as a function of pulse length Δt_{p}, along with the deduced fitting parameters. (b) Sketch aiding the derivation of the heating transient. (c) Exemplary dependence of conversion efficiency $\eta_{1\,\mathrm{kW}}$ on thermal impedance Z_{th}, emulating the behavior in CW-mode at the respective thermal resistance R_{th}.

was introduced as thermal impedance $Z_{\mathrm{th}}(\Delta t_{\mathrm{p}}) = \Delta T_{\mathrm{AZ}}(\Delta t_{\mathrm{p}})/P_{\mathrm{diss}}$. The developed model describes the data well, with the fit parameters $R_{\mathrm{th}} \approx 0.05$ K/W, $\tau_{\mathrm{th}} \approx 2$ ms extracted. For comparison purposes, the theoretical CW-value of the probed material is estimated based on the thermal conductivity values λ_{th}. For a bar consisting of n_{e} emitters, each of width w, and the cavity length L, the CW thermal resistance is estimated as

$$R_{\mathrm{th}} = \frac{1}{L\, n_{\mathrm{e}}\, w} \cdot \sum_k \frac{d_k}{\lambda_{\mathrm{th},k}}, \tag{4.1}$$

with d_k the thickness of the k-th layer and its thermal conductivity $\lambda_{\mathrm{th},k}$. This assumes one-directional heat flow, vertically through the layer stack, and neglects potentially present boundary resistances. Likewise, the thermal capacity is given as

$$C_{\mathrm{th}} = L\, n_{\mathrm{e}}\, w \cdot \sum_k d_k\, \varrho_k\, c_{\mathrm{th},k}, \tag{4.2}$$

introducing ϱ_k the mass density and $c_{\mathrm{th},k}$ the specific heat capacity of the k-th layer. Table 4.3 lists the individual contributions for the specific case of the laser bars studied herein, with all layers between the active zone, i.e. a region of a few microns thickness in which most of the heat is generated [Rie18], and the solder right underneath the p-

Layer	$R_{th,k}$ / K W^{-1}	Fraction / %	$C_{th,k}$ / J K^{-1}	Fraction / %
Chip	$2.3 \cdot 10^{-3}$	4.5	$1.2 \cdot 10^{-4}$	0.3
Metallization	$4.1 \cdot 10^{-4}$	0.8	$4.1 \cdot 10^{-4}$	1.1
Solder	$2.1 \cdot 10^{-3}$	4.3	$4.7 \cdot 10^{-4}$	1.3
Submount	$4.0 \cdot 10^{-2}$	81.8	$3.5 \cdot 10^{-2}$	94.5
Solder	$4.2 \cdot 10^{-3}$	8.6	$1.1 \cdot 10^{-3}$	2.8
Sum	0.05	100	0.037	100

Table 4.3: Theoretical derivation of thermal resistance R_{th} and capacity C_{th} based on the thermal conductivity and capacity values of the processed material stack. Material parameters taken from [Pip13, Smi76].

side submount. With 82 and 95%, the submount contributes by far the most to the two quantities. The estimated theoretical thermal resistance and time constant are

$$R_{th}^{theo} \approx 0.05\,\mathrm{K/W}, \qquad \tau_{th}^{theo} = R_{th}^{theo} \cdot C_{th}^{theo} \approx 1.9\,\mathrm{ms}.$$

Firstly, this good match justifies the simplifications made in order to get an idea of the temporal behavior one can expect from a certain device design. Secondly, the analysis shows that heating in this device is restricted mainly to the region down to the submount during the time ranges probed here.
As an application of this method, Figure 4.4(c) exemplarily depicts the conversion efficiency $\eta_{1\,\mathrm{kW}}$ at 1 kW as a function of the thermal impedance Z_{th}. For the very bar design presented here, the efficiency slowly falls with increasing impedance, but stays within the 1%-point-error-margin of the test station (cf. Sec. 3.2.2).

One way to interpret this graph is the following: If this chip design is mounted on advanced high-performance coolers, exhibiting a thermal resistance shown on the graph's abscissa, and is operated in CW-mode, it will experience the same heating as in these QCW-experiments (at the respective thermal impedance) and may consequently operate at the depicted respective efficiency. In short, duty cycle-dependent QCW-testing with passive cooling can be used to emulate the behavior expected in CW-operation with high-performance active cooling[12]. This method then saves costly high-performance-mounting when scanning through different bar designs.

[12]High-performance cooling in CW-operation with thermal resistance as low as [0.05:0.06] K/W has been reported in [Kna11, Str18].

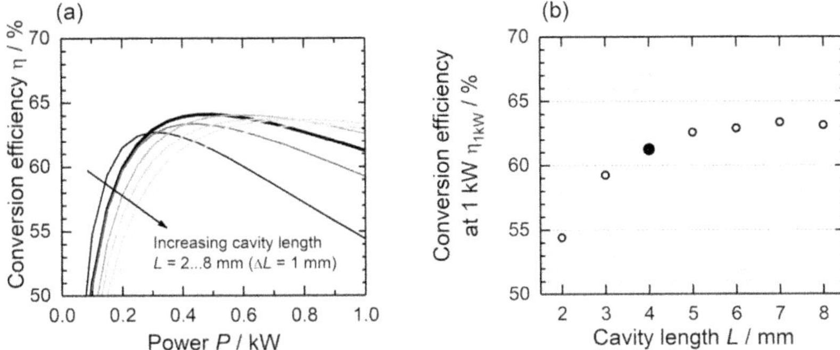

Figure 4.5: Design study for increased 1 kW-efficiency via varied cavity length L. (a) Based on measurement results from a fabricated 4 mm-long bar, the performance of bars with cavity lengths between 2 and 8 mm are extrapolated. (b) From this extrapolation, the conversion efficiency at 1 kW is plotted here as a function of cavity length.

4.3. Varied resonator length

This section analyzes the potential for performance improvements by variation of the longitudinal cavity length L. The goal behind is to reduce the series resistance

$$R_s = \frac{\rho_a}{L\, n_e\, w} \propto 1/L, \tag{4.3}$$

with ρ_a the areal series resistance, L the cavity length, n_e the number of emitters, and w the emitter width. Lengthening the cavity thus appears suitable for reducing the operation voltage $V(I) = V_0 + I \cdot R_s$, thereby increasing the conversion efficiency [Kar17, Str18].

As a baseline, 4 mm-long devices with the lateral structure $n_e \cdot w = 37 \cdot 186\,\mu\text{m}$ were fabricated from the vertical structure studied in Section 4.1. Based on its lasing parameters acquired in low-thermal tests[13], such as threshold current I_{th}, series resistance R_s, slope efficiency S, internal loss α_i, and modal gain Γg_0, the performance of a laser bar with varied cavity length L was extrapolated. The results can be viewed in Figure 4.5, with lengths between 2 and 8 mm considered, scanned in 1 mm-steps.

For this extrapolation, the series resistance was simply scaled via the dependence above,

[13]QCW-test-conditions: pulse length $\Delta t_p = 200\,\mu\text{s}$, repetition rate $f_{rep} = 10$ Hz.

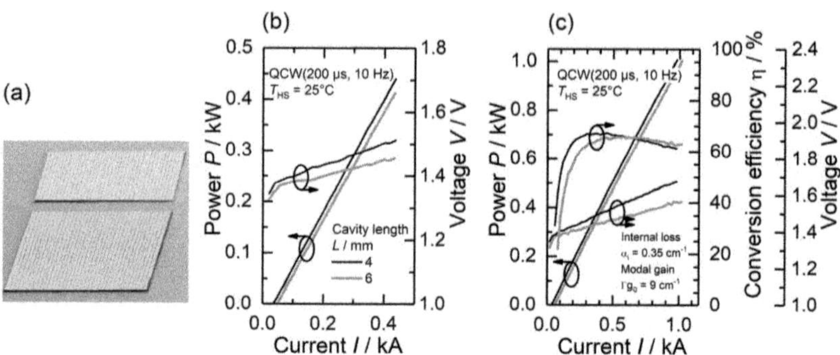

Figure 4.6: Comparison of fabricated bars with cavities of 4 and 6 mm length. (a) Fully processed chips with 37 emitters of 186 μm width (©FBH/schurian.com). (b) Optoelectronic characterization of these designs at low currents up to 400 A and low-heating conditions. (c) The same designs now measured up to 1 kW at the same QCW conditions.

while the threshold current

$$I_{\text{th}} \propto L \cdot \exp\left(-\frac{\ln(R_{\text{f}}R_{\text{r}})}{2L\Gamma g_0}\right) \tag{4.4}$$

depends twofold on the cavity length: Firstly, longer cavities increase the active volume that needs to be pumped and thus increase the threshold. Secondly, a mode traveling along the resonator is amplified more strongly in a longer cavity, reducing the threshold. The slope efficiency was scaled for varied length L via

$$S \propto \left(1 - \frac{2L\alpha_{\text{i}}}{\ln(R_{\text{f}}R_{\text{r}})}\right)^{-1}, \tag{4.5}$$

highlighting the importance of low internal loss α_{i}. In both cases, the rear-facet reflectivity was held at $R_{\text{r}} = 98\%$ and the front-facet reflectivity adjusted for best performance $R_{\text{f}} \in [0.5 : 3]\%$. Hereby, longer cavities need to couple out a higher fraction of the emission, e.g. $R_{\text{f}} \approx 0.5\%$ at $L = 8$ mm.

While the 4 mm-long baseline design emits 1 kW with 61% conversion efficiency, shorter cavities suffer from the increased series resistance reducing the efficiency to 59 and 54%, for cavity length 3 and 2 mm, respectively. On the contrary, lengthening the resonator indeed increases the conversion efficiency. For 6 mm-long devices the efficiency increases by 2%-points to 63%. Using even longer chips based on this vertical structure is, however, not worthwhile as the efficiency saturates. Although the series resistance continues to fall,

Cavity length L / mm	**4**	**6**	Difference	(Expected)
Front-facet reflectivity R_f / %	2	1		
Series resistance R_s / mΩ	0.33	0.23	-30%	(-33%)
Threshold current I_{th} / A	36	47	+33%	(+30%)
Slope efficiency S / WA^{-1}	1.1	1.08	-2%	(-2%)
Efficiency at 1 kW η(1 kW) / %				
QCW(200 μs, 10 Hz)	61	63	+2%-pts.	(+2%-pts.)

Table 4.4: Experimental performance of designs with the two cavity lengths L = 4 and 6 mm, along with the expected changes from Equations 4.3 - 4.5.

the slope efficiency is compromised too much for further efficiency increases. Essentially, even longer cavities suffer from the (in part unavoidable) optical loss α_i present in the vertical structure.

Based on this extrapolation, 6 mm-long bars were fabricated and then tested in low-thermal conditions (Δt_p = 200 μs, f_{rep} = 10 Hz). Figure 4.6 compares its performance to the 4 mm-long reference. The graph in panel (a) tests the basic operation parameters at comparatively low currents, i.e. if the design goals were realized as fabricated. Indeed, the series resistance is reduced by a third, as is also listed in Table 4.4. Likewise, the threshold current increases by roughly a third, which is the penalty of lengthening the resonator. Due to the rather low internal loss α_i = 0.35 cm^{-1} of the vertical structure used, the slope efficiency is compromised by only 2%. When measured up to the operation point P = 1 kW, the longer-cavity design shows the previously extrapolated efficiency increase to 63% [Kar17]. For all currents I > 600 A this improved structure operates at a higher efficiency than the 4 mm-long baseline.

The observed performance improvement is crucially tied to the low internal loss α_i = 0.35 cm^{-1} afforded by the used vertical structure. To illustrate, Figure 4.7 shows the performance of bar designs fabricated from a different vertical structure optimized for high modal gain Γg_0, but compromising the internal loss with α_i = 0.55 cm^{-1}. Consequently, the slope efficiency is reduced by an unacceptable margin when lengthening the resonator, cf. Eq. 4.5, with the effect that for this vertical structure the 6 mm-long device does not

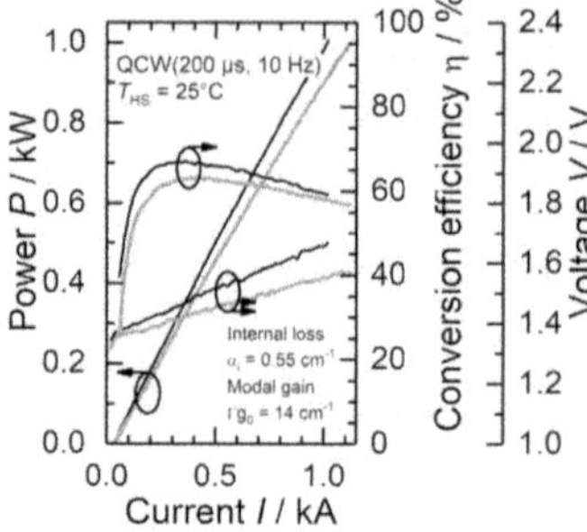

Figure 4.7: Optoelectronic performance at varied cavity length for a vertical structure with compromised internal loss. The conversion efficiency of the long-resonator bar (gray) never exceeds that of the shorter device.

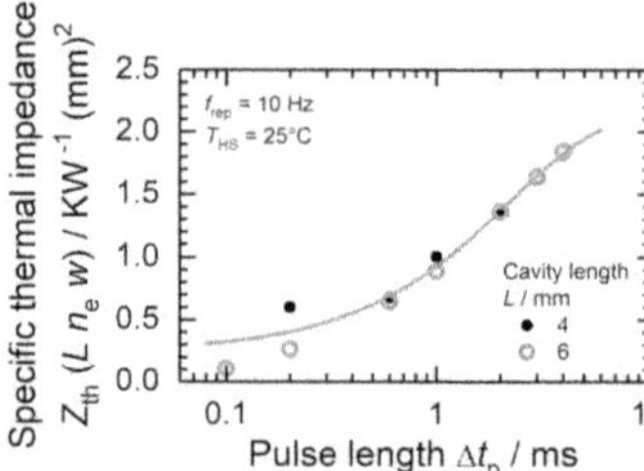

Figure 4.8: Pulse length-dependent specific thermal impedance measured for two cavity lengths, along with a fit to the joint dependence.

outperform its 4 mm-long counterpart at any current value.
Once again, using the low-loss vertical structure, the efficiency at 1 kW is increased when lengthening the resonator to $L = 6\,\mathrm{mm}$. This improvement is now tested at higher thermal load using the transient method described in the previous Section 4.2. The conversion efficiency maintains its high value $\eta = 63\%$ at higher thermal loads with pulse lengths up to $\Delta t_\mathrm{p} = 2\,\mathrm{ms}$ ($f_\mathrm{rep} = 10\,\mathrm{Hz}$). The data was already shown in Figure 4.4(c) when illustrating the thermal-transient method. The inferred thermal impedance $Z_\mathrm{th} = 0.033\,\mathrm{K/W}$ at the longest pulses translates to a heating of

$$\Delta T_\mathrm{AZ} = Z_\mathrm{th} \cdot P \cdot (1/\eta - 1) \approx 19\,\mathrm{K}.$$

Further, the larger-footprint bar benefits from improved cooling through the larger base area. In turn, at equal QCW conditions Δt_p, f_rep, the long-cavity device exhibits reduced thermal impedance compared to its short-cavity counterpart. This dependence is illustrated in Figure 4.8, where a specific thermal impedance is defined by scaling with the active base area $L\ n_\mathrm{e}\ w$. The values inferred from both cavity lengths follow a joint dependence. Comparing to Equations 4.1 and 4.2,

$$R_{\text{th}} \cdot (L\ n_{\text{e}}\ w) = \sum_k \frac{d_k}{\lambda_{\text{th},k}}, \qquad \tau_{\text{th}} = \left(\sum_k \frac{d_k}{\lambda_{\text{th},k}}\right) \cdot \left(\sum_k d_k \varrho_k c_{\text{th},k}\right)$$

both the convergence value $R_{\text{th}} \cdot (L\ n_{\text{e}}\ w)$ and the time constant τ_{th} solely depend on the layer sequence in the device, but not on, for instance, the cavity length L, under the assumptions made.

4.4. Impact of the lateral design

The efficiency of diode-laser bars operating at the operation point 1 kW was found to be limited by the series resistance R_{s} [Sch07, Kar17]. An alternative route to lengthening the resonator, which was discussed in Section 4.3, is to increase the total emitter width $n_{\text{e}} \cdot w$. The choices in emitter width w and number of emitters n_{e} are limited by two technological boundary conditions. Firstly, the spacing d_{s} between adjacent emitters cannot fall below a certain value ($d_{\text{s}} \geq 65\,\mu\text{m}$ in this work) in order to leave sufficient space for required etch processes. Secondly, the mechanical bar width must not exceed 10 mm, with the aperture

$$w_{\text{ap}} = (w + d_{\text{s}})(n_{\text{e}} - 1) + w$$

preferably only ≈ 9.2mm to ensure safe handling and mounting of the bar chip. The lateral layouts possible with these conditions noted are displayed in Figure 4.9(a) along with the resultant total emitter width $n_{\text{e}} \cdot w$. Comparing to the baseline lateral layout ($n_{\text{e}} = 37$, $w = 186\,\mu\text{m}$, cf. Sec. 4.1), $n_{\text{e}} \cdot w$ can only be increased (and thus R_{s} reduced) if fewer but wider emitters are used. Following this reasoning, Table 4.5 lists three different lateral layouts A, B, C of ascending total emitter width $n_{\text{e}} \cdot w$ that were designed and processed as 4 mm-long bars. The selected emitter widths range from 186 to 1095 μm, with the wider emitters requiring sub-structuring via a shallow implantation step [Spr10, Kar17a, Cru17]. This prevents ring modes from experiencing sufficient gain, otherwise posing a significant loss mechanism [Pit01, Pod19]. Figure 4.9(b) estimates the projected increase in efficiency for the two high-fill-factor layouts B and C. This estimation uses the slope efficiency S, threshold current I_{th}, voltage offset V_0, and series resistance R_{s} seen from the baseline structure ($37 \times 186\,\mu\text{m}$), and then extrapolates via scaling to higher fill-factor

$$I_{\text{th}} \propto n_{\text{e}} \cdot w, \quad R_{\text{s}} \propto \frac{1}{n_{\text{e}} \cdot w},$$

which approximately holds if $\Delta x_{\text{spread}}/w \ll 1$, cf. Sec. 2.2. In this estimate, the efficiency at the optical power $P = 1$ kW is improved from 61% to 62 and 63% for the two high-

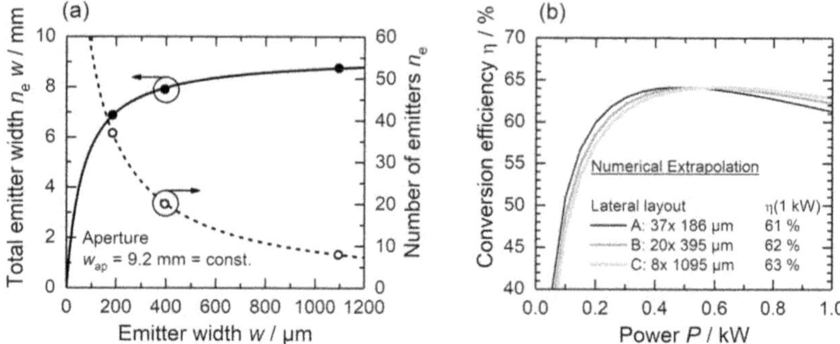

Figure 4.9: Design study for reduced series resistance via increased total emitter width. (a) For constant aperture $w_{ap} = 9.2$ mm and emitter spacing $d_s = 65\,\mu$m lateral layouts can only fall on the dashed curve of combinations of emitter width w and number of emitters n_e. (b) Estimated conversion efficiency for three selected lateral layouts based on the experimentally determined characteristic of layout A exhibiting the lowest total emitter width $n_e \cdot w$.

fill-factor layouts, respectively. The expected increase in threshold current is thus more than compensated by the reduction in series resistance, as a result of the required high operation current.

Measurement results of these three structures, recorded at low duty cycle 0.2% (pulse width $\Delta t_p = 200\,\mu$s, rep. rate $f_{rep} = 10$ Hz) are shown in Figure 4.10(a) and the deduced values of threshold current, series resistance, and slope efficiency are listed in Table 4.5. As expected, the threshold current increases proportionally with $n_e \cdot w$. Likewise, the series resistance is reduced at higher fill-factor, while the low-current slope efficiency is maintained. Exceeding the estimates made above, the efficiency at the operation point is even increased to 63 and 66% [Kar19, Kar20, Cru21], respectively. The root for this behavior is illustrated in Figure 4.10(b), in which the slope efficiency $S = \mathrm{d}P/\mathrm{d}I$ is plotted as a function of emission power P. While all three layouts start off at the same low-current value $S \approx 1.2$ W/A, the two high-fill-factor layouts show significantly reduced roll-over. At the operation point $P = 1$ kW, the baseline structure rolls over to ≈ 0.9 W/A, while layouts B and C still show ≈ 1.1 W/A.

Next, the structures are investigated regarding their behavior at elevated heating conditions arising from increased duty cycle. Results of this duty-cycle series are shown exemplarily for layout C with 87% fill-factor in Figures 4.11(a-b). As seen in panel (a), the behavior hardly changes as the duty cycle is increased. Panel (b) illustrates how the heating in the active zone is tracked via the change in emission wavelength upon opera-

Lateral layout	**A**	**B**	**C**
Emitter width w / μm	186	395	1095
Number of emitters n_e	37	20	8
Emitter spacing d_s / μm		65	
Total emitter width $n_e \cdot w$ / mm	6.9	7.9	8.7
Fill-factor FF / %	69	79	87
Emitter sub-structuring	No	Yes	Yes
Threshold current I_{th} / A	46	54	57
Series resistance R_s / mΩ	0.32	0.30	0.27
Slope efficiency $S(I \approx I_{th})$ / WA^{-1}	1.17	1.16	1.17
Efficiency at 1 kW η(1 kW) / %			
QCW(200 μs, 10 Hz), (i.e. $R_{th} \approx 0.02$ K/W)	61	63	66
QCW(2 ms, 20 Hz), (i.e. $R_{th} \approx 0.05$ K/W)	54	61	64

Table 4.5: Design and performance summary of three lateral layouts for high-efficiency 1 kW-emission. The maximum efficiency 66% is seen from layout C at low-heating conditions.

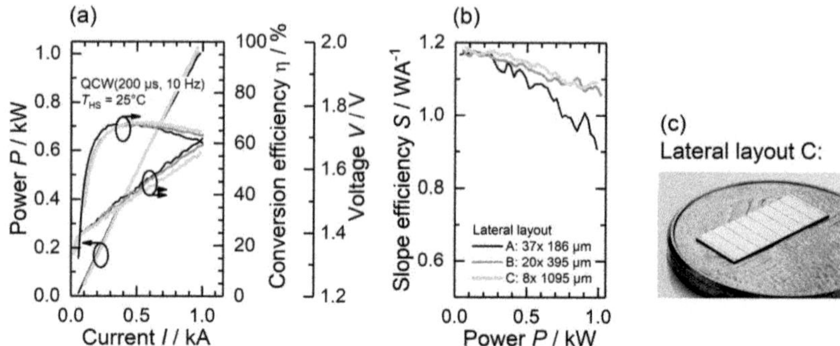

Figure 4.10: Measurement results of three lateral high-efficiency designs acquired at low-heating conditions, duty cycle 0.2%. (a) Output power P, conversion efficiency η, and voltage V as a function of operation current I. (b) Deduced slope efficiency S as a function of emission power P. (c) Fully processed bar chip exhibiting eight 1095 μm-wide emitters (©FBH/schurian.com).

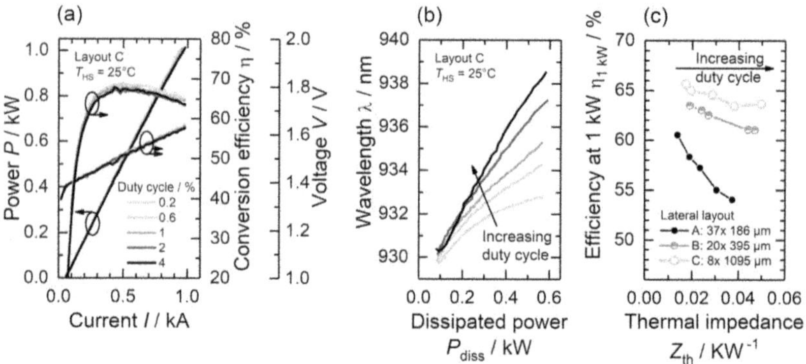

Figure 4.11: The three lateral high-efficiency designs were investigated at increasing duty cycle [0.2:4]%. (a) Power-voltage-current curves exemplarily shown for layout C. (b) Emission wavelength λ as a function of dissipation power P_{diss} exemplarily shown for layout C, used to infer the heating level for each operation point and duty cycle. (c) Efficiency at 1 kW emission power at increased duty cycle, i.e. thermal impedance Z_{th}, for the three lateral layouts A, B, and C.

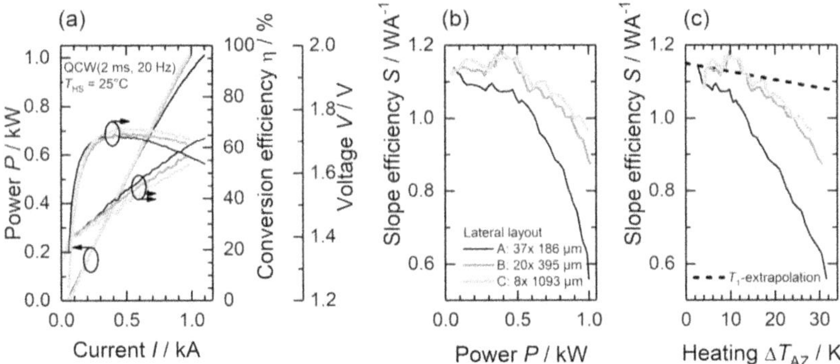

Figure 4.12: Measurement results of three lateral high-efficiency designs acquired at high-heating conditions (2 ms, 20 Hz, i.e. $Z_{\mathrm{th}} \approx 0.05$ K/W). (a) Power-voltage-current characteristic. (b) Deduced slope efficiency as a function of emission power. (c) Slope efficiency re-plotted as a function of heating, along with the upper roll-over limit from T_1-extrapolation.

tion, as discussed in Section 4.2. A higher duty cycle generates a higher time-averaged dissipation power and thus heats up the structure more strongly, manifesting in a stronger shift in wavelength. Again, the different heating conditions at varied duty cycle can be expressed as varying thermal impedance Z_{th}, herein ranging between $\approx [0.02 : 0.05]$ K/W. In order to visualize the performance at varying heating, Figure 4.11(c) tracks the efficiency at 1 kW with increasing thermal impedance for all three lateral layouts. Not only does layout A exhibit the lowest efficiency of the three layouts at low heating, but its efficiency also decreases the fastest with increasing thermal impedance, down to 54% at highest heating $Z_{\mathrm{th}} \approx 0.05$ K/W. A lower decrease, by 2%-points, is observed for layouts B and C, yielding still 61 and 64% [Kar20], respectively. The behavior at highest heating (2 ms-pulses at 20 Hz rep. rate) is studied in more depth in Figure 4.12. The realized design goal of reduced series resistance is clearly visible in the $I-V$-curve in panel (a). Even more pronounced in this high-heating case is the difference in roll-over, as highlighted in panel (b). Now, layout A rolls over to ≈ 0.56 W/A at 1 kW, while the other two layouts still show ≈ 0.97 W/A. This improved roll-over characteristic is the main contributor to the overall efficiency benefit. In order to understand this effect, panel (c) re-plots the slope efficiency of panel (b) now as a function of heating

$$\Delta T_{\mathrm{AZ}}(P) = Z_{\mathrm{th}} \cdot P_{\mathrm{diss}}(P)$$

at the same QCW conditions 2 ms, 20 Hz. This shows that the different roll-over characteristics do not simply stem from different heating levels in the three layouts, e.g. as a result of the different series resistances. Instead, the three curves still do not lie on top of each other. This does, however, indicate that one of the physical mechanisms responsible for roll-over is mitigated when using wider stripes $w \geq 400\,\mu$m. A potential mechanism that drives roll-over in response to heating and may further be emitter width-dependent is described in [Rau17]. Here, the lateral temperature profile of each emitter created upon operation acts as a lateral waveguide confining the lateral modes to a region smaller than the provided lateral current distribution. This leaves gain at the emitter edges unused, manifesting in ever-reduced slope efficiency. This edge effect has a smaller impact in wider emitters, thus extracting the total provided gain more effectively [Kar20]. Panel (c) also shows the upper limit for reduced roll-over by plotting the T_1-based slope efficiency (cf. Eq. 2.4)[14]. As the experimentally determined slope efficiency for the wide emitters still falls below this upper limit, at least one more roll-over mechanism must be present, of which longitudinal spatial hole burning is perceived to dominate by [Kau19].

4.5. Conclusions

This chapter studied how to develop highly-efficient 1 kW-emitting laser bars operating at room temperature. Different design approaches were first tested numerically against their potential to increase the efficiency at the operation point and then implemented as processed devices to verify their benefit.

Firstly, a suitable vertical structure was introduced with the concept of an EDAS (extreme double-asymmetric) large optical cavity at its foundation. It was chosen for its per-design low optical loss and low series resistance, which were both verified in processed devices. The internal loss measured as low as $\alpha_i = 0.35\,\mathrm{cm}^{-1}$. The series resistance was determined to $R_s = 0.33\,\mathrm{m}\Omega$ for a 4 mm-long bar with 37 emitters of 186 μm width, corresponding to an areal series resistance of $\rho_a = R_s \cdot A \approx 9\,\mathrm{m}\Omega\ (\mathrm{mm})^2$. This baseline design emits the targeted power with an efficiency of 61% at low-thermal conditions, i.e. thermal resistance $R_{th} \approx 0.02\,\mathrm{K/W}$.

The temporal behavior of the lasing emission was then studied on the millisecond-scale by varying the QCW duty cycle. The assessed range of pulse lengths spanned $\Delta t_p \in [0.1 : 4]$ ms. Tracking the lasing wavelength as a function of the dissipated power at each operation point allowed the computing of the heating ΔT_{AZ} in the device. The retrieved pulse

[14] $T_1 \approx 600\,\mathrm{K}$ [Kau19]

length-dependent thermal resistance, termed thermal impedance Z_{th}, followed a non-linear function of the duty cycle. Rate equation-based modeling allowed for the extraction of the CW-value of the probed volume (0.05 K/W for a 6 mm-long bar) and the characteristic time constant $\tau_{th} = 2$ ms. Both values were reproduced by theoretical estimates based on the material constants of the layer sequence. As to the practical use, by varying the duty cycle, and thus the thermal impedance, one can study the behavior of a particular bar design at various thermal conditions. In one extreme, low-duty-cycle measurements reveal the upper-limit efficiency of a certain design, reachable in CW operation only with ever-improving mounting and cooling technology. In the other extreme, CW operation with advanced state-of-the-art technology can be emulated using a higher duty cycle. This is possible since the passively-cooled mounting technology used in this thesis shows, at high duty cycle, thermal impedance values similar to the thermal resistance of advanced actively-cooled CW-operation heatsinks $R_{th} \approx [0.05 : 0.06]$ K/W.

The following two studies sought to increase the conversion efficiency η further by reducing the series resistance. With the vertical structure maintained and the chip width 1 cm being a set device property one is left with lengthening the resonator or increasing the bar's fill-factor FF. Pursuit of the first resulted in an efficiency increase by 2%-points to 63% at the resonator length $L = 6$ mm (low-thermal testing). The improvement is attributed to (i) a reduction of the series resistance by a third, and of equal importance, to (ii) hardly compromised slope efficiency S (−2%) thanks to the low optical loss α_i of the used vertical structure. Performance extrapolation to even longer cavities showed plateauing efficiency of 63% despite a further reduction in series resistance, while suffering from increasingly compromised slope efficiency. Future developments for such resonator lengths therefore would require even lower optical loss in order to realize performance gains. A maximum of 65% was extrapolated for $L = 8$ mm at (to date unrealistically low) optical loss $\alpha_i = 0.1\,\text{cm}^{-1}$ and all other parameters kept. The further experimental study of the 6 mm-long devices showed the reduction of the thermal resistance R_{th} from its 4 mm-value, which helped to maintain the efficiency $\eta = 63\%$ at the operation point even at elevated thermal conditions up to $R_{th} = 0.035$ K/W (QCW 2 ms, 10 Hz).

Even higher efficiency was obtained when using wider emitters instead of lengthening the resonator. Lateral layouts comprising emitters of widths 395 and 1095 μm were investigated. Along with wider emitters, the fill-factor increased from 69 to 79 and 87%, respectively. The benefit of a larger footprint manifested in a reduction in series resistance, decreasing from 0.32 to 0.30 and 0.27 mΩ, overcompensating for the increase in threshold current from 46 to 54 and 57 A. The slope efficiency at low currents maintains $S \approx 1.17$ W/A. In effect, the layout with 1095 μm-wide emitters sets the new efficiency benchmark of 66% at 1 kW, in low-heating conditions ($R_{th} \approx 0.02$ K/W). The particular

Description	**Design**	η**(1 kW) / %**
Baseline	4 mm, 37 × 186 μm	61
Long cavity	6 mm, 37 × 186 μm	63
High fill-factor	4 mm, 8 × 1095 μm	66
Long cavity and high fill-factor	6 mm, 8 × 1095 μm	68 projected
+ latest vertical structure [Ars19]	6 mm, 8 × 1095 μm	$\geq$ 70 projected

Table 4.6: Summary of the efficiency values at 1 kW demonstrated in this chapter. Here, all test conditions were QCW(200 μs, 10 Hz) for $R_{\mathrm{th}} \approx 0.02$ K/W. In addition, two feasible near-term goals are projected.

benefit of wide emitters is clearly visible at elevated thermal load ($R_{\mathrm{th}} \approx 0.05$ K/W). While the baseline structure ($w = 186\,\mu$m) rolls over notably ($S(1\,\mathrm{kW}) \approx 0.56$ W/A), the record design shows reduced power saturation with $S(1\,\mathrm{kW}) \approx 1$ W/A and thus still an efficiency of 64%. This is a value also feasible for CW conditions if the chip is mounted into highly-advanced state-of-the-art cooling architectures with similar thermal resistance. More effective extraction of the provided gain is held accountable for the roll-over mitigation in the wide-emitter layout. The edge effect of a laterally compressed mode upon chip heating loses its strong influence on deteriorated gain extraction the wider the emitter.

The demonstrated efficiency advances $\eta(1\,\mathrm{kW})$, using essentially two different technological measures, allow projecting feasible near-term goals, cf. Tab. 4.6. For once, lengthening the longitudinal resonator and increasing the fill-factor may easily be combined in one chip, such that the efficiency of 68% may be reached. In addition to that, the utilization of the latest vertical structure (and its benefit on power saturation mechanisms such as longitudinal spatial hole burning) should allow efficiency values exceeding 70% at 1 kW emitted power.

5. Efficient 1 kW-Emitting Bars Exhibiting Narrow Far Field

In this chapter, the physical mechanisms governing the beam quality of high-power laser bars are studied. Again targeting the operation point of 1 kW emitted power, the developed bar designs make use of findings from the previous chapter in order to ensure efficient operation. The first section examines the effect of bonding-induced chip deformation, commonly termed "bar smile", on the lateral emission divergence. In particular, two bar designs, one providing lateral confinement via index-guiding trenches and one relying on gain-guiding, are contrasted with respect to their resistance to bonding-associated stress fields. Measurements are carried out at a low duty cycle of 0.2% as to not let the studied effects be overshadowed by thermal influences. The next two sections, on the other hand, are devoted to thermal effects. In an exemplary exercise, the distance between adjacent emitters is varied in an attempt to shape the lateral temperature profile across each individual emitter. Emitter-resolved far-field acquisition then allows extracting the resultant effect onto the beam quality. Lastly, the technology of emitter sub-structuring is explored for bars comprised of wide emitters $w = 1095\,\mu$m. For this, each bar emitter is composed of a number of sub-emitters defined by periodic implantation of the contact layer. The measurement results in this chapter are, where suitable, complemented by numerical simulations.

5.1. Mechanical chip deformation (bar smile)

Mechanical deformation of bar chips post mounting, commonly referred to as "bar smile", is a well known phenomenon. Soldering intrinsically strained bar chips, themselves comprising of layers with different thermal expansion coefficients, onto non-ideally planar submounts of yet another expansion coefficient leaves so mounted diode lasers in a bent shape. As detailed in Section 3.2.4, the smile is commonly quantified as the height of this catenary.

The following examines the effect of non-vanishing smile onto the lateral far-field width and contrasts a baseline design with a new bar development that aims to beneficially affect the found correlation. This study of laterally varying strain fields $\varepsilon_{kl}(x)$, $k, l \in \{x, y\}$ uses comparably narrow emitter stripes $w = 186\,\mu$m in order to provide high spatial resolution. In the baseline design [Fre16, Kar17a, Fre19], each bar emitter is surrounded by dry-etched index-guiding (IG) trenches as detailed in Section 3.1 acting as $\Delta n_{\text{eff}} = 10^{-3}$. The new development omits etched trenches and relies on gain guiding (GG). It applies the deep implantation technique instead, with parameters (depth, offset from stripe edge) chosen for a comparable current path as in the design with IG trenches. Bars from both

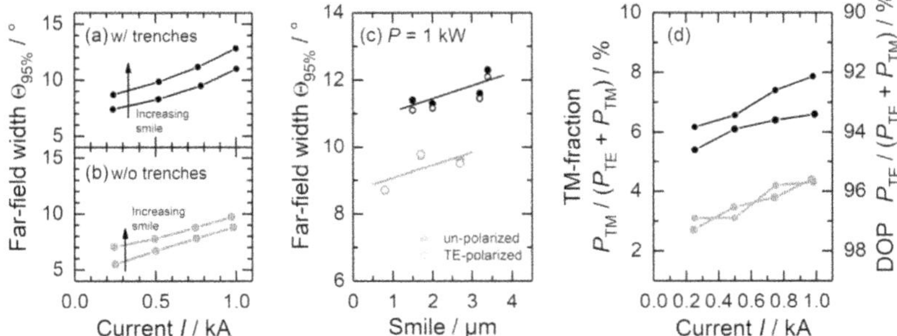

Figure 5.1: Effect of bar smile onto the lateral far field. (a) Overall bar far-field width $\Theta_{95\%}$ as a function of injection current I for IG-based bars with varied smile. (b) Same dependence of bars without etched trenches relying on gain guiding. (c) Far-field width at 1 kW as a function of the bar smile for the two bar technologies. (d) Fraction of transverse-magnetic (TM) emission at increasing current I.

designs were mounted such that a range of smile values is observable and its effect can be studied. Measurements were carried out in low-thermal conditions (QCW 200 μs, 10 Hz, $R_{\text{th}} \approx 0.02\,\text{K/W}$) such that the stress-induced effects can be studied in an isolated manner. The deduced bar far-field width $\Theta_{95\%}$ is displayed in Figure 5.1(a,b) as a function of the operation point, with the new gain-guiding design showing reduced divergence across all currents [Kar19]. Panel (c) extracts the far-field width at $P = 1\,\text{kW}$ which rises with increasing bar smile, but with a $\approx 2°$-reduction from designs with etched trenches to those without. The graph further compares the width of the total (unpolarized) to the transverse-electric (TE) polarized emission, where small reductions in the IG trench case and virtually no differences in the GG-case are visible. This is also in line with the lower fraction of transverse magnetic (TM) polarized emission in the GG-bars, cf. panel (d), although it rises with increasing current. In consequence, the degree of polarization (DOP) for these nominally TE-emitting bars

$$\text{DOP} = \frac{P_{\text{TE}}}{P_{\text{TE}} + P_{\text{TM}}}$$

ranges between [96:97]% and [92:94]% for GG and IG-bars, respectively.

The following seeks to identify the physical mechanism through which the high smile-sensitivity in trench-featuring bars is mediated and thus examines three probable root causes by turning to an in-detail study of one IG-device with high smile $\approx 3\,\mu$m.

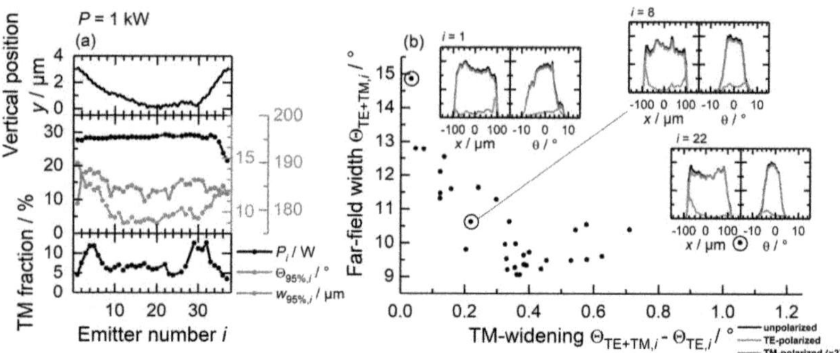

Figure 5.2: In-depth study of exemplary 3 μm-smile bar using IG trenches. (a) Emitter-resolved presentation of vertical position (bar smile), power P_i, far-field width $\Theta_{95\%,i}$, near-field width $w_{95\%,i}$, and TM-fraction. (b) Far-field width of the total (unpolarized) emitter emission as a function of the far-field widening due to TM-contributions with insets of three exemplary emitters displaying polarization-resolved near- and far-field profiles.

- **TM-addition:**

The smile profile of the studied bar is displayed in the topmost panel of Figure 5.2(a), complemented by emitter-resolved data on power-per-emitter P_i, emitter far-field width $\Theta_{95\%,i}$, and near-field width $w_{95\%,i}$, as well as the fraction of TM-polarized emission. Interestingly, while the emitter-power distribution is comparably flat, the far-field width distribution follows the smile curve in shape with low values in the bar center and widened far field at the bar edges. The TM-fraction follows this trend only in parts as it increases towards the bar edges but then again falls to values also seen in the bar center. This already suggests that the pronounced far-field width variation of the unpolarized emission is not simply due to smile-induced TM-contributions and panel (b) examines this more closely. The emitter-resolved far-field examination was here carried out for the total unpolarized emission, the TE-polarized, and the TM-polarized emission separately. The deduced widths $\Theta_{\mathrm{TE+TM},i}$, $\Theta_{\mathrm{TE},i}$, and $\Theta_{\mathrm{TM},i}$, respectively are the base for this graph, in which the unpolarized width $\Theta_{\mathrm{TE+TM},i}$ (ranging between $\approx 9°$ and $15°$) is plotted as a function of far-field widening owing to TM-contributions (at maximum $1.1°$ corresponding to 12%). If the TM-contributions were accountable for the observed behavior one would find a proportional relation, but the contrary is the case. Those emitters with widest far field exhibit the lowest TM-widening and vice versa. Insets with polarization-resolved near- and far-field profiles of three exemplary emitters illustrate the little deviation between

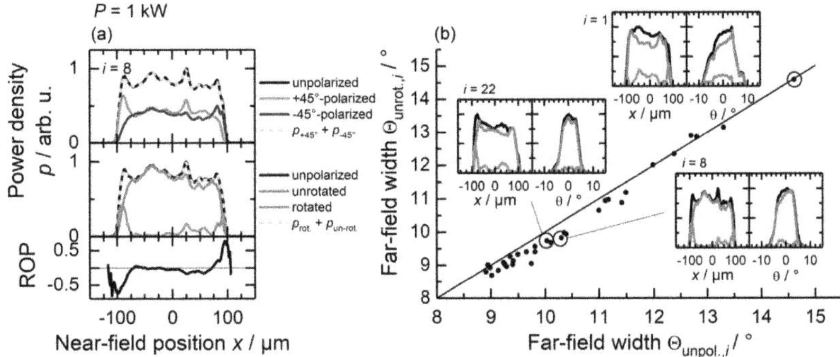

Figure 5.3: Continued in-depth study of the emitter-resolved polarization state, now in a reference system rotated by 45°. (a) Near-field profiles from an exemplary emitter without polarizer and when filtered at two different polarizer angles, followed by the deduced distributions of rotated and unrotated power density. The bottom panel illustrates magnitude and orientation of the polarization-state rotation through the ROP. (b) Far-field width of the unrotated emission as a function of the unpolarized (total) far-field width compared to a bisecting diagonal (slope of unity, vanishing intercept). Three insets show exemplary near- and far-field profiles resolved with respect to polarization rotation.

unpolarized (black) and TE-polarized (green) distributions. Thus the plain addition of wide TM emission to the overall far field is not accountable for the strain-induced far-field widening – in contrast to the reasoning in [Win14].

- **TM-rotation:**

With this physical mechanism ruled out, another probable cause is examined. Several studies [Bie07, Cas13a, Cas13] observed a rotation of the polarization axis when strained devices were studied in PL experiments, cf. Sec. 2.3. This rotation occurs since the nominally cubic semiconductor crystal becomes birefringent when subject to shear strain $\varepsilon_{xy} \neq 0$. It is thus not unlikely that TM-contributions, typically being wider than TE emission [Win14], may be rotated whilst traveling inside stressed emitters leading to considerable projections onto the TE-axis and thus the observed widening even in the TE-direction. To test this thesis, the experimental approach in above PL-based publications is here transferred to active devices and expanded by separating rotated and unrotated emission contributions.

The experimental procedure is the following: Emitter-resolved near- and far-field profiles $p(x)$, $p(\theta)$ are acquired again but now for emission, which is (i) unpolarized, (ii) passing the polarizer when rotated to +45° to the emitter base plane, and (iii) polarized along −45°, as illustrated for an exemplary emitter in the topmost panel of Figure 5.3(a). The idea is that if any emission is rotated by not more than ±45° from its original direction[15], which appears justified according to the conclusions in Section 2.3, then subtracting the profiles acquired in +45° and −45° from each other leaves those power contributions having experienced polarization rotation, weighted by their corresponding angle.

$$p_{\text{rotated}} := |p_{-45^\circ} - p_{+45^\circ}| \tag{5.1}$$

This reasoning becomes immediately clear when considering the two limiting cases: (i) pure TE-emission projects equally onto the +45° and −45° axes ($p_{+45^\circ} = p_{-45^\circ}$) leaving no rotated contributions, (ii) emission rotated by +45° from its original axis is fully accounted for as rotated power since $p_{+45^\circ} = 1$, $p_{-45^\circ} = 0$ and thus $p_{\text{rotated}} = 1$. The unrotated contributions are derived via

$$p_{\text{unrotated}} := (p_{+45^\circ} + p_{-45^\circ}) - p_{\text{rotated}} = p_{+45^\circ} + p_{-45^\circ} - |p_{-45^\circ} - p_{+45^\circ}|. \tag{5.2}$$

Exemplary results are shown in the middle panel of Figure 5.3(a) where most rotated contributions are found at the emitter edges. When defining the rotated degree of polarization (ROP)

$$\text{ROP} := \frac{p_{-45^\circ} - p_{+45^\circ}}{p_{+45^\circ} + p_{-45^\circ}}, \tag{5.3}$$

of which an exemplary distribution is shown in the bottom panel of Figure 5.3(a), one can infer the anti-symmetric distribution of the polarization rotation and thus of the shear strain along the emitter width. The same rotation-dependent analysis was carried out for all other emitters examined in near- and far-field configuration. Exemplary data is shown in the insets in Figure 5.3(b), illustrating magnitude and distribution of rotated (red) and unrotated (green) contributions. Interestingly, most rotated power density is found rather around the center of the far-field profiles. The main graph in panel (b) plots the derived far-field width of the unrotated contributions $\Theta_{\text{unrot.},i}$ as a function of the width $\Theta_{\text{unpol.},i}$ of the total (unpolarized) emission for the same 3 μm-smile IG-bar studied above. The emitters either fall onto or below the bisecting diagonal indicating that polarization rotation does widen the unpolarized far field in most cases. Its magnitude ($< 0.7^\circ$) does, however, not only not account for the observed smile-induced far-field widening but also is small for wide emitters and vice versa. Thus, the proposed mechanism of wide TM emission being rotated onto the TE axis is not accountable for strain-induced far-field widening.

[15]The original polarization axes are 0° and 90° for heavy- and light-hole emission, respectively

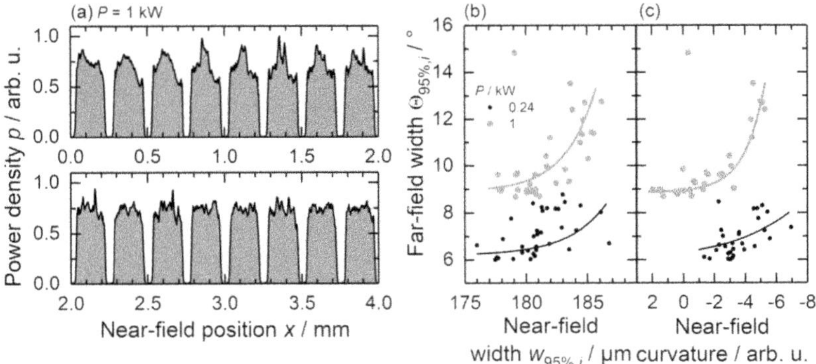

Figure 5.4: Determination of the root cause for smile-induced far-field widening. (a) Near-field profiles of the first eight (top) and the following eight emitters (bottom) reaching towards the bar center. (b) Individual emitter far-field widths $\Theta_{95\%,i}$ as a function of the corresponding near-field width at two different operation points. (c) Far-field width plotted against the near-field profiles' curvature.

- **Waveguiding:**

The governing root cause is identified when comparing the near-field shapes along the bar, cf. Fig. 5.4(a). The upper panel displays the first eight emitters while the lower one shows the following eight emitters reaching towards the bar center. While those center emitters reveal a close-to-top-hat shape with a nearly flat cap, the closer the emitter to the bar edge the more structured the near field becomes. Pronounced contributions in the emitter center and sharply falling flanks start occurring indicating that the emission is shaped by lateral waveguides. Intermittently occurring sharp power-density spikes indicate narrow local confinement. When quantifying waveguide-induced shaping via the near-field width $w_{95\%,i}$ and the profile curvature, for which parabolic functions were fitted to the distributions, clear correlations to the far-field width $\Theta_{95\%,i}$ are inferred, cf. Fig. 5.4(b,c). Both, rising near-field width and curvature lead to a non-linear increase in far-field width. Excepted from this is only the edge emitter with the widest far field whose near-field width and curvature fall below the trend.

Now the response of bars with and without IG trenches to bar smile is investigated and for this the above-studied IG trench bar is compared to a bar of the new development mounted to exhibit a virtually identical smile profile, cf. Fig. 5.5(a). While the bar with

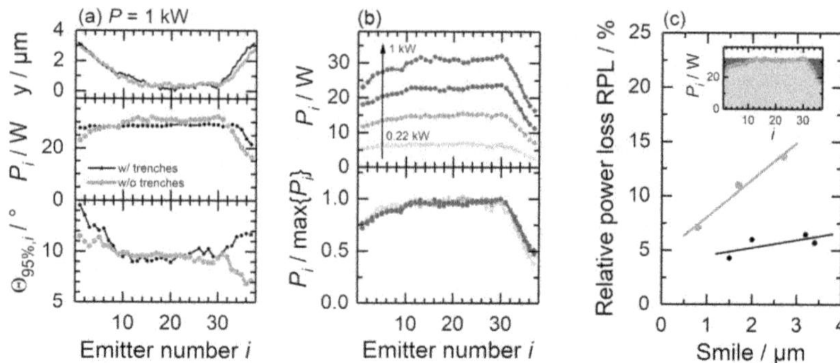

Figure 5.5: Response of designs with and without etched trenches to bar smile. (a) Comparison of the two technologies, mounted with the same smile profile (top panel), with respect to emitter power (middle) and far-field width (bottom) distributions. (b) Emitter-power distributions at rising operation point (top) and normalized to the respective maximum (bottom). (c) From emitter-power distributions deduced RPL for bars of the two technologies and mounted with different smile.

trenches shows a mostly flat emitter-power distribution (middle panel), the one from the bar without trenches is prominently curved with steadily decreasing power towards the bar edges. The far-field width distribution (bottom) of this high-smile bar using the new development reveals less prominent increases towards the bar edges than were observed for the above in-detail studied trench-featuring bar of the same smile. Thus the new bar technology without index-guiding trenches is less sensitive to non-vanishing smile with respect to far-field widening. This is attributed to the fact that trenches act as stress accumulators in a stressed chip, cf. Fig. 2.3(a).

The characteristic observed in the emitter-power distribution is studied in more detail in Figure 5.5(b), where data is shown for four different operation points (top). The normalized curves (bottom) fall on top of each other indicating that the accountable physical mechanism does not depend on the absolute operation point but is of relative nature. Smile-induced polarization-axis rotation in structures where there is little to no optical gain for TM contributions does act as a relative loss mechanism and does explain the seen behavior. For a given strain and thus photo-elastically-induced birefringence distribution, a constant fraction of the traveling power is rotated regardless of the absolute power level[16]. Rotated contributions do, however, project onto the TM axis and require

[16]Dependence on the operation point may, however, be present when tested at high heating (e.g. CW) where the strain field may rearrange as a result of different thermal expansion.

TM-gain in order to not be absorbed along the cavity and to be emitted from the facet[17]. In fact, the trench-omitting design shows lower TM power, cf. Fig. 5.1(d) and thus appears to be prone to this "power frown". The IG design, on the other hand, accumulates stress at its trenches, cf. Fig. 2.3(a), which alters the band structure in this region [Bie04], and may provide sufficient TM gain to support the rotated power and finally lead to reduced power frown and reduced DOP of the device, as observed.

A quantitative comparison of this behavior is supported by the definition of the *relative power loss* (RPL). It estimates the fraction of power that is lost due to the proposed mechanism and assumes that at least one emitter shows full power[18] $\max\{P_i\}$ without losses and computes the power $\max\{P_i\}-P_i$ lost in every emitter referring to this maximum power. This sum is then normalized by the assumed total power when no losses were present and all n_e emitters radiated this maximum power.

$$\mathrm{RPL} := \frac{\sum_{i=1}^{n_\mathrm{e}}(\max\{P_i\} - P_i)}{n_\mathrm{e} \cdot \max\{P_i\}} = 1 - \frac{P}{n_\mathrm{e} \cdot \max\{P_i\}} \tag{5.4}$$

The inset of Figure 5.5(c) gives a graphical representation of this definition as the ratio of the dark gray area to the sum of light and dark gray areas. The RPL comparison of bars mounted with different smile and of the two technologies in this figure shows that (i) the RPL increases with smile and that (ii) this increase is stronger in bars without trenches, supposedly due to lower gain for TM contributions.

The beneficial effect of the new technology without IG trenches onto the far field is now exploited in a bar mounted to yield non-vanishing but low smile $\leq 0.8\,\mu$m. Its far-field width is shown in Figure 5.6(a) as a function of emission power and at 1 kW it observes [Kar19, Kar20]

$$\Theta_{95\%}(1\,\mathrm{kW}) = 8.8°.$$

As a result of the flat smile profile, cf. Fig. 5.6(b, top), both emitter-power and far-field width distribution (bottom) are rather flat, too. The stacked[19] near-field profile shown in panel (c) is approximately of top-hat shape and 183 μm wide. The far-field profile in panel (d) reveals sharp flanks with small tails.

[17] As elaborated on in Section 2.3, [Hol18] assumed complete conversion of any TM photon absorbed into an TE photon via relaxation into the heavy-hole band. Apparently, this is not the case, but induces optical losses and this heating, which may explain the reduced reliability in (externally) stressed devices quoted by many publications.

[18] This also implies that the RPL is underestimated in high-smile bars where there is no emitter left not subject to strain-induced losses.

[19] Numerical superposition of all emitters' near-field profiles emulating the use of routinely applied beam rotators as, e.g. in [Haa17].

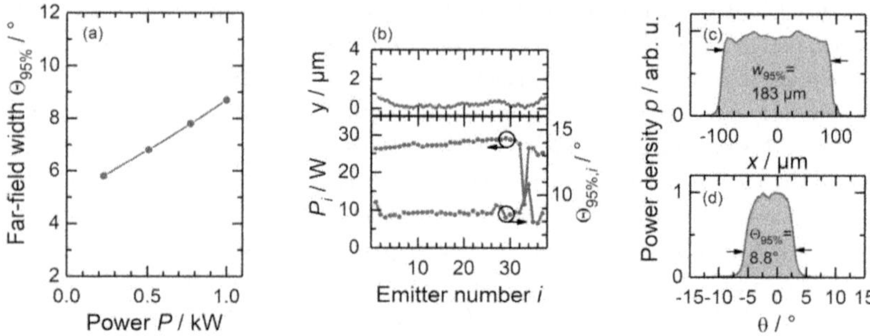

Figure 5.6: Highest brightness from the new bar development obtained at QCW conditions (200 μs, 10 Hz, $R_{\mathrm{th}} \approx 0.02$ K/W) (a) Far-field width as a function of emission power. (b) Bar-smile profile (top) complemented by emitter-power and far-field width distribution at 1 kW (bottom). Super-imposed near-field profile (c) and far-field distribution (d) at the operation point along with the respective width.

5.2. Emitter cross heating

This section aims to investigate the influence of thermal emitter cross-talk on the far-field width. The underlying principle is based on the observation (cf. Sec. 2.2) that each diode-laser emitter builds up a lateral temperature profile upon operation as a result of anisotropic heat generation and its inhomogeneous conduction towards the heat sink. The increasing curvature of this temperature (thus reflective index) profile at increasing operation point is then a substantial contributor to the widening of the far field.

Thermal emitter cross-talk shall now flatten each emitter's temperature profile and thus reduce the far-field width of the emitted light [Kar20, Cru22]. For this exemplary exercise, two bar layouts were designed and processed that essentially differ in the distance d_{s} in-between adjacent emitters. This comparison is built around the deduced characteristic thermal length $d_{\mathrm{th}} \approx 240\,\mu$m (cf. Sec. 2.4) and thus it is expected that for

- $d_{\mathrm{s,A}} = 255\,\mu\mathrm{m} > d_{\mathrm{th}}$ emitters operate thermally individually,
- $d_{\mathrm{s,B}} = 65\,\mu\mathrm{m} < d_{\mathrm{th}}$ emitters experience thermal cross-talk.

The cavity measures $L = 4$ mm in length and all emitters are processed nominally identically with a sub-structure ($w_{\mathrm{sub}} = 20\,\mu$m, $p_{\mathrm{sub}} = 29\,\mu$m) as described in Section 3.1 and 5.4. No index-guiding trenches were etched beneath the stripes.

In a first step, the bars were characterized at low heating in QCW-mode (pulse length $\Delta t_{\mathrm{p}} = 200\,\mu$s, rep. rate $f_{\mathrm{rep}} = 10$ Hz) at which a thermal resistance $R_{\mathrm{th}} \approx 0.02$ K/W

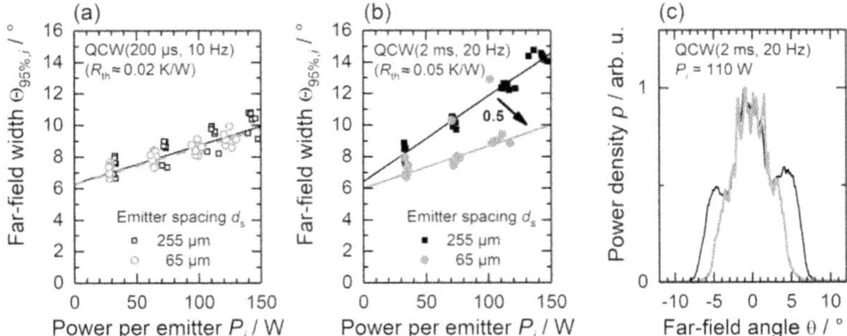

Figure 5.7: The effect of reduced emitter spacing d_s on the lateral far field, analyzed on individual-emitter level. (a) Dependence of the far-field width on the power per emitter at low duty cycle. (b) Same dependence, now measured at elevated duty cycle with corresponding linear fits (only non-edge emitters in case d_s = 65 μm). (c) Far-field profiles of individual bar emitters (both same power level) at elevated duty cycle, comparing high (black) and low (gray) emitter spacing.

Lateral layout	**A**	**B**
Emitter spacing d_s / μm	255	65
Number of emitters n_e	7	8
Emitter width w / μm	1095	
Far-field degradation $\frac{d\Theta}{dP}$ / ° W^{-1}		
QCW(200 μs, 10 Hz), (i.e. $R_{th} \approx 0.02$ K/W)	0.027	0.027
QCW(2 ms, 20 Hz), (i.e. $R_{th} \approx 0.05$ K/W)	0.054	0.027

Table 5.1: The effect of reduced emitter spacing d_s on the lateral far field, analyzed on individual-emitter level. The listed degradation slopes originate from the linear fits shown in Fig. 5.7(a, b).

was inferred. Figure 5.7(a) plots the far-field width $\Theta_{95\%,i}$ of the individual bar emitters as a function of their respective emission power P_i. At these test conditions, the two structures appear degenerate as they follow the joint dependence of 6° intercept angle and $\approx 0.027°/\text{W}$ far-field widening at increasing power. The similar behaviors suggest that thermal cross-talk between the emitters is negligible and hence does not affect the mode shapes and composition. This changes at elevated-heating test conditions ($\Delta t_\text{p} = 2$ ms, $f_\text{rep} = 20$ Hz) for which the far-field width is again plotted against the individual emitter power in Figure 5.7(b). While the offset angle 6° is unchanged (indicating a purely thermal effect) the widening rate is doubled to $\approx 0.054°/\text{W}$ for the emitters spaced at $d_\text{s,A} = 255\ \mu\text{m}$. Compared to the low-thermal case, the emitters now seem to be subject to a laterally varying temperature profile that confines and shapes modes with a wider far field. The seven emitters processed on this bar behave mutually comparably and thus operate thermally individually. This supports the estimated characteristic thermal length $d_\text{th} \approx 240\ \mu\text{m} < d_\text{s,A}$ as the minimally required spacing for non-overlapping temperature profiles at the tested thermal resistance. The emitters exhibit a far-field width of $\Theta_{95\%,i} \approx 14°$ at $P_i = 144$ W. Half the widening slope, and thus the same dependence as in the low-thermal case, is revealed by the non-edge emitters of layout B with the spacing $d_\text{s,B} = 65\ \mu\text{m}$. Following the dependence

$$\Theta_{95\%,i} = 6° + 0.027°/\text{W} \cdot P_i, \tag{5.5}$$

these emitters appear to benefit from the reduced emitter spacing d_s. Its decrease from 250 to 65 μm reduced the widening slope by 50% as summarized in Table 5.1. The one outer emitter at each bar end does not follow the dependence of all other inner emitters - an effect which is studied in more detail in the following Section 5.3. Figure 5.7(c) highlights the improved far-field characteristic by comparing the far-field profiles for the two spacing values at the same emission power $P_i = 110$ W. While the profiles are very similar for all angles ±3°, they differ prominently at higher angles. The profile of the structure with lower spacing then steadily decays to zero while the other structure reveals an extra shoulder symmetrically at each side being accountable for the strongly widened far field.

After the experimental verification of the beneficial effect of nearer emitters it is now linked more closely to the presumed underlying cause – the thermal emitter cross-talk. For this, the temperature distribution across these two types of bar structures was simulated[20], for which results are displayed in Figure 5.8(a) chosen for identical power per emitter $P_i = 125$ W which is an eighth of 1 kW. This operation point corresponds to

[20]Kindly, thermal simulations were performed by Dipl.-Ing. S. Grützner (TRUMPF Laser Berlin) with the COMSOL multiphysics software based on steady-state finite-element analysis of a given material- and heat-source distribution. Details of the assumptions made were presented in [Rie18].

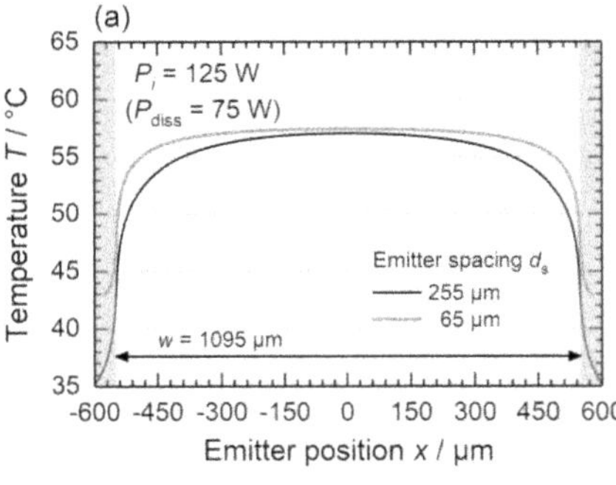

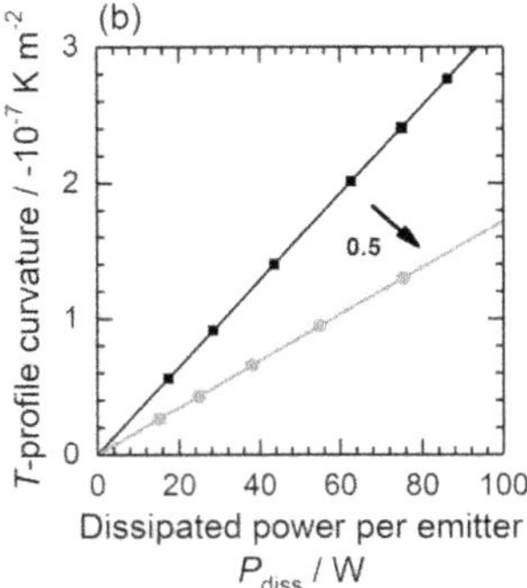

Figure 5.8: The simulated effect of reduced emitter spacing on the temperature distribution across the emitters. (a) Lateral temperature profiles for the same operation point $P_i = 125$W (b) Curvature extracted from the simulated temperature profiles as dependent on the operation point.

$P_{\text{diss}} = 75$ W dissipated power per emitter. While both profiles peak at the same chip temperature 57°C they do reveal distinct differences in shape. The dependence for higher spacing steadily falls from its maximum at the emitter center and reaches the temperature 44°C at the emitter edges. Its minimum is observed between adjacent emitters amounting to 35°C. The profile for reduced spacing instead shows a less curved dependence which further leads to an increased emitter-edge temperature 48°C and minimum point 43°C. This flattening is a result of reduced lateral heat flux and thus edge cooling due to the close adjacent emitter's temperature gradient. These qualitative differences shall now be quantified for the further analysis. [Win16, Win18] developed a phenomenological measure of a temperature profile's impact on the resultant far field. It was found that in broad-area lasers the profile's curvature, deduced via parabolic fits, increases linearly with the measured far-field width. The curvature then coarsely quantifies the thermally-induced waveguide building up upon operation. This observation was reproduced and made use of in a variety of designs by [Rie18, Rie19]. Following this convenient method, Figure 5.8(b) plots the curvature derived from parabolic fits to the profiles in (a) as a function of the individual-emitter operation point. Aside from the absolute curvature values – now a measure of the expected far-field width – it is rather its relative increase that is of concern when comparing the two lateral layouts. It is found that the curvature increases with half the slope when the emitter spacing is reduced from 255 to 65 μm. This is in line with and can hence be attributed to the experimentally observed reduction in far-field degradation by 50%, cf. Fig. 5.7(b) and Table 5.1.

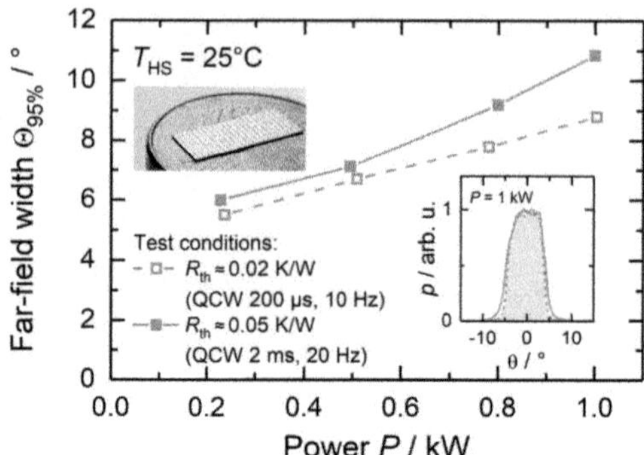

Figure 5.9: Results from high-brightness bar design exploiting thermal emitter cross-talk. The design features 186 μm-wide emitters at the low spacing $d_s = 65\ \mu$m and is tested at two heating conditions. The insets display the far-field profiles at the two thermal conditions and the processed chip (©FBH/schurian.com).

Reducing the spacing d_s between adjacent emitters from 255 to 65μm was successfully implemented to narrow the far field of the individual bar emitters. While this demonstration was exercised using a stripe width $w = 1095\ \mu$m its basic optimization concept can likely be generalized to other stripe widths, in particular those used in this thesis $w = 186$ and 400 μm. To illustrate the benefit of the insights gained so far, the following bar structure was designed: (i) low emitter distance $d_s = 65\ \mu$m for beneficial emitter cross-heating using high-brightness $w = 186\,\mu$m-emitters, (ii) the deep implantation technique for reduced in-chip strain fields as compared to index-guiding-trench design, (iii) low-smile bonding $\leq 0.8\ \mu$m for reduced external stress source. The results are displayed in Figure 5.9 acquired at the two heating cases $R_{th} \approx 0.02$ and 0.05 K/W corresponding to the QCW test conditions (200 μs, 10 Hz) and (2 ms, 20 Hz), respectively[21]. At 1 kW emitted power the bar design exhibits a lateral divergence of 8.8 and 10.8° [Kar20], respectively.

5.3. Bar edge emitters

Another consequence of the beneficial thermal cross-talk between neighboring emitters, studied in the last section, arises from the finite bar chip width. While all inner emitters are subject to the generated heat from the respective emitter to the left and to the right

[21]low-heating results repeated from Sec. 5.1 for convenience

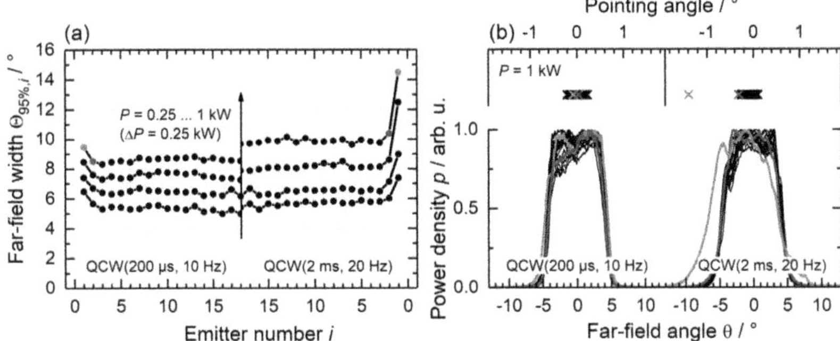

Figure 5.10: Effect of the bar edge on the individual emitters' far field. In both panels, results at low and elevated heating are compared on the left and right, respectively. (a) Far-field width of the individual bar emitters at increasing bar emission power P. (b) The individual emitters' far-field profiles plotted on top of each other with the two outmost emitters highlighted in color. Above, on a separate enlarged scale a visualization of the respective lateral pointing angles.

(local line symmetry), emitters at the bar edge experience a non-symmetric heat-flux field. This may result in a non-symmetric temperature distribution across these outer emitters if they are placed at a distance $d_s < d_{th}$ such that they do not operate thermally individually [Cru22]. It is now investigated how this affects the far-field pattern of the emitted light. The following results were obtained from a bar with a lateral layout comprising of $n_e = 37$ emitters of $w = 186\ \mu m$ width placed with a spacing $d_s = 65\ \mu m$. No index-guiding trenches were etched beneath the stripes. The bar was first tested at low heating level ($\Delta t_p = 200\ \mu s$, $f_{rep} = 10$ Hz) corresponding to a thermal resistance $R_{th} \approx 0.02$ K/W, followed by measurements at high heating ($\Delta t_p = 2$ ms, $f_{rep} = 20$ Hz) for a thermal resistance also reached by modern cooling architectures for CW operation $R_{th} \approx 0.05$K/W.

Figure 5.10(a) depicts the 95%-power-content far-field width $\Theta_{95\%,i}$ of the individual emitters on the bar chip, compared at the two heating levels in the panel left and right, respectively. For clarity, only one half of the bar (18 emitters) is depicted, as the results are line-symmetric around the chip width center. Starting right from low emission powers (250 W), the outmost emitter $i = 1$ reveals a wider far field than all other ones, while no further systematic deviation along the residual bar is apparent. At 1 kW bar power, the far-field width's mean value of all emitters $i \neq 1$ amounts to 8.6° from which the outmost emitter $i = 1$ (red) sets itself apart by $\approx 1°$. The second emitter's (blue) far field is not significantly widened above average. At higher heating, the outmost emitter (red) stands

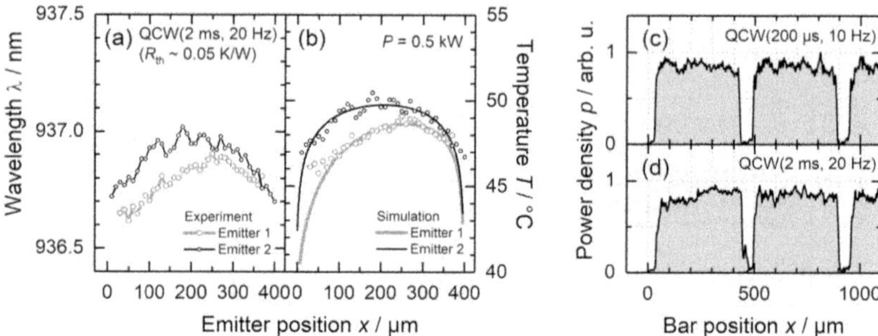

Figure 5.11: Lateral temperature profiles as origin of the observed far-field widening at the bar edge. (a) Central wavelength along the emitter width measured at elevated heating (b) Lateral temperature profile as computed from the data in (a) along with the simulated profiles. (c) Near-field profile of the emitters displayed in (a, b) measured at low duty cycle revealing alike top-hat shapes. (d) Near-field profiles of the same emitters, now at elevated heating, showing an asymmetric shape for the first emitter.

out even more prominently with its width increased by 4.6° above the average value 9.8° at 1 kW. While the emitter second to the bar edge (blue) reveals a slight increase in far-field width (0.5° above average at 1 kW), the third and all other emitters are again not affected by this edge effect.

Figure 5.10(b) shows the far-field profiles of all the emitters corresponding to the data shown in panel (a). Again, the low-heating result is on the left and shows a family of similar profiles with none of them particularly standing out. The center-of-mass angles (a measure of the lateral emission pointing) plotted right above the profiles with an extra, enlarged scale hover around the forward direction 0°. In the right panel, at elevated heating, all emitters but the first two outer ones ($i \neq 1, 2$) appear similar, while the outer emitters ($i = 1, 2$) do not only reveal a widened far-field pattern but also a pointing deviating from the forward direction by up to 1.4°. The two outer emitters on one bar edge shine to the left and the outer emitters on the other bar edge to the right (line anti-symmetry). With regard to the overall bar far field $\Theta_{95\%}$, this thermally-induced pointing contributes power at high angles where all other emitters' far fields have decayed leading to further increases in $\Theta_{95\%}$.

While this effect was now studied exemplarily in a bar using 186 μm-wide emitters, it was similarly found in bars using $w = 395\,\mu$m- and in those using $w = 1095\,\mu$m-wide emitters (again, all emitters identical per bar and omission of index-guiding trenches) with the same spacing $d_s = 65\,\mu$m. In case of these emitter widths, the edge perturbation affects only

Stripe width w/μm	Widening of bar far field $\Delta\Theta_{95\%}(1\,\mathbf{kW})$ / °
186	0.5
395	1.1$^{\#}$
1095	3.0

Table 5.2: Contribution of the edge-emitter high-angle intensity to the overall bar far field. The bar far field was calculated from the individual emitter profiles in- and excluding the two outmost emitters with the obtained difference listed here. Elevated heating ($\Delta t_\mathrm{p} = 2$ ms, $f_\mathrm{rep} = 20$ Hz) $^{\#}$at 0.8 kW.

the outmost emitters on each bar edge $i = 1, n_\mathrm{e}$. For the elevated-heating case, Table 5.2 lists the increase in bar far-field width $\Theta_{95\%}$ resultant from this edge effect as a function of the stripe width w.

Figures 5.11(a, b) illustrates the origin of the inferred far-field pointing and widening. In the example of now $w = 395\,\mu$m, the lateral temperature profile across an edge and a non-edge emitter is plotted as a function of the lateral position x. Experimental results were deduced from measurements of the position-resolved emission wavelength, cf. panel (a). For this, the near-field pattern was scanned with an opening of 10 μm and the collected light analyzed regarding its wavelength spectrum. While the second emitter shows a line-symmetric wavelength distribution, the profile of the edge emitter is clearly asymmetric. It is now tested if these wavelength variations correspond to the temperature differences expected in this structure. Absolute temperatures are computed from the measured data noting the structure heating $\Delta T_\mathrm{AZ} = R_\mathrm{th}\, P_\mathrm{diss}$ above heat sink temperature T_HS, and local variations are attributed to refractive index-driven wavelength shifts of the longitudinal mode. Appendix B details the derivation and estimates a change rate $(\mathrm{d}\lambda/\mathrm{d}T)_\mathrm{longit.} \approx 0.085$ nm/K. The local wavelength $\lambda(x)$ is then converted to the local temperature $T(x)$ via

$$T(x) = T_\mathrm{HS} + R_\mathrm{th}\; P\; (1/\eta - 1) + (\lambda(x) - \lambda_0)\; / \; \left(\frac{\mathrm{d}\lambda}{\mathrm{d}T}\right)_\mathrm{longit.}$$

using the listed parameters[22]. The computed temperature profiles in (b) reveal line symmetry for the second emitter $i = 2$ and a non-symmetric shape for the edge emitter $i = 1$. These experimental results are supported by thermal simulations plotted along with the experimental data. The mutual agreement supports the assumption that the longitudinal modes, with their conversion factor $(\mathrm{d}\lambda/\mathrm{d}T)_\mathrm{longit.} \approx 0.085$ nm/K, give rise to the inferred local wavelength variations.

This deduced temperature distribution of the semiconductor material leads to a laterally varying refractive index across the emitters. Thus a further guiding mechanism for

[22] $T_\mathrm{HS} = 25°\mathrm{C}$, $R_\mathrm{th} = 0.05$ K/W, $P = 0.5$ kW, $\eta = 64\%$, $\lambda_0 = 936.05$ nm

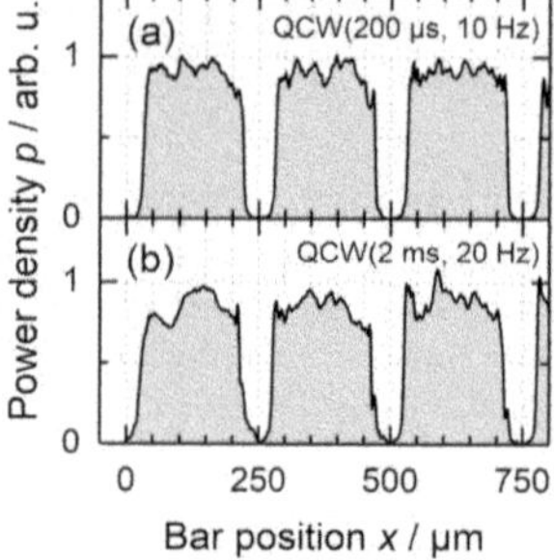

Figure 5.12: Thermal bar-edge effect as visible in $w = 186\,\mu$m-bar near-field profiles, acquired at the two heating conditions.

the supported lateral modes comes into play in addition to the non-thermal gain guiding case. The change in refractive index between emitter center and edge is estimated[23] as $\Delta n_{\mathrm{r}} \approx 10^{-3}$ and thus of a magnitude with significant impact on the guided modes.
Its effect can be inspected in Figure 5.11(c, d) where the resultant near-field profile, covering the first two to three emitters, is plotted. At low heating ($R_{\mathrm{th}} \approx 0.02\,\mathrm{K/W}$), the first two emitters $i = 1, 2$ exhibit a line-symmetric top-hat pattern. A potentially present temperature distribution apparently has insufficient influence on the gain-guided modes in both emitters. This changes at elevated heating, since while emitter $i = 2$ still shows local line symmetry, the edge-emitter profile is clearly compressed on the bar-edge side (here, left side) with less steep edges of the near-field pattern. The temperature (and thus refractive index) variation has reached a substantial level to visibly shape the supported modes. This asymmetric near-field shaping then plausibly explains the lateral pointing of this emitter's far field inferred above. For completeness, the same shaping is observed in the corresponding near-field profiles of the $186\,\mu$m-wide emitters studied at the beginning of this section, cf. Fig. 5.12.

It was found that the discontinuity of the heat flux field at the bar edge disturbs the otherwise periodic lateral temperature profile $T(x) = T(x + w + d_{\mathrm{s}})$. For the lateral layout $w = 186\,\mu\mathrm{m}$, $d_{\mathrm{s}} = 65\,\mu\mathrm{m}$, this discontinuity decays up to a point within the second emitter (within the first emitter for $w = 395, 1095\,\mu\mathrm{m}$) and hence within a lateral length that compares well to the characteristic thermal length $d_{\mathrm{th}} \approx 240\,\mu\mathrm{m}$ estimated earlier. This decay length thus appears as a suitable predictor that only the first emitter is affected strongly, as

$$w + d_{\mathrm{s}} > d_{\mathrm{th}}, \forall\ w \text{ investigated herein.}$$

[23] $\Delta n_{\mathrm{r}} \approx \Delta T \cdot \mathrm{d}n_{\mathrm{r}}/\mathrm{d}T$, with $\Delta T \approx 3\,\mathrm{K}$, $\mathrm{d}n_{\mathrm{r}}/\mathrm{d}T = 3.2 \cdot 10^{-4}\,\mathrm{K}^{-1}$ for GaAs at $\lambda = 940\,\mathrm{nm}$ [Ska03]

All other inner emitters lie beyond this decay length, are exposed to the periodic temperature profile $T(x) = T(x + w + d_s)$ and thus exhibit alike far-field profiles.
While the studied edge-emitter effect appears as detrimental – which in itself it unarguably is – it should be kept in mind that this effect originates from the overall beneficial thermal emitter cross-talk all inner bar emitters profit from.

5.4. Emitter sub-structure

The studies on efficiency improvements via optimized lateral design, cf. Sec. 4.4, illustrated the suitability of wide emitters $w = 1095\,\mu$m, which is why a closer look on the resultant beam characteristic is worthwhile. For proper operation of such unconventional broad-area lasers, the emitters must not be contacted electrically across the entire emitter area. [Pit01] observed the onset of ring modes (also visualized in [Pod19]) even in 200 μm-wide fully-contacted emitters which led to partially undirected emission and poor efficiency. To prevent the amplification of light traveling laterally, current injection is impeded periodically via shallow implantation of narrow longitudinal stripes into the contact layer following [Spr10, Pla14]. This concept is herein applied to laser bars and first bar-based emission profiles are shown.
The studied bars exhibit seven emitters spaced at a distance $d_s = 255\,\mu$m for thermally individual operation. Each $w = 1095\,\mu$m-wide emitter is sub-structured into sub-stripes (electrically conducting area between adjacent implantation stripes) of width $w_{\text{sub}} = 20\,\mu$m pitched at $p_{\text{sub}} = 29\,\mu$m and thus spaced with 9 μm in-between adjacent sub-stripes.

The retrieved far-field width is plotted against the emission power in Figure 5.13(a). At low duty cycle DC=0.2% ($R_{\text{th}} \approx 0.02$ K/W) the 95%-width amounts to 10.4° at 1 kW emitted power, while the profile has widened to 14.5° at high duty cycle DC=4% ($R_{\text{th}} \approx 0.05$ K/W). The 2.5-fold increase in thermal resistance is reflected by the far-field degradation slope $\mathrm{d}\Theta/\mathrm{d}P$. It increases from 3.0°/kW to 7.4°/kW by a factor of 2.5.
The following results reveal the instructive competition between coherent coupling and individualized operation of these sub-emitters. The far-field patterns emitted by the bar are shown in Figure 5.13(b) for the low-duty-cycle case at four operation points. At about 5 W emitted power the profile is composed of two narrow spikes indicating a so-called array super-mode [Agr85]. Interaction between adjacent sub-stripes with a phase shift of $\Delta\phi = \pi$ experiences low optical loss [Bot86] and thus the angle θ for constructive interference can be estimated via

$$\theta \approx \pm \arcsin\left(\frac{\lambda}{2p_{\text{sub}}}\right) = \pm 0.93^\circ$$

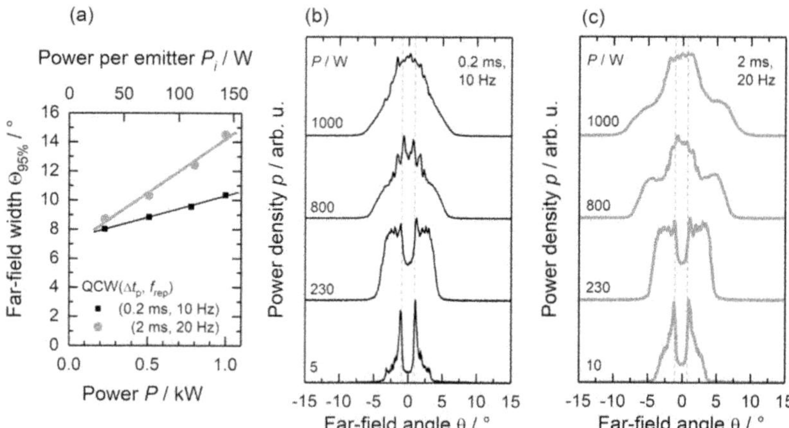

Figure 5.13: Far-field characteristic of bars with sub-structured 1095 μm-wide emitters. (a) Far-field width as a function of emission power. (b) Far-field profiles at increasing operation point measured at low duty cycle. The angles $\pm 0.93°$ are marked with dashed lines. (c) The same, now acquired at high duty cycle and thus elevated heating.

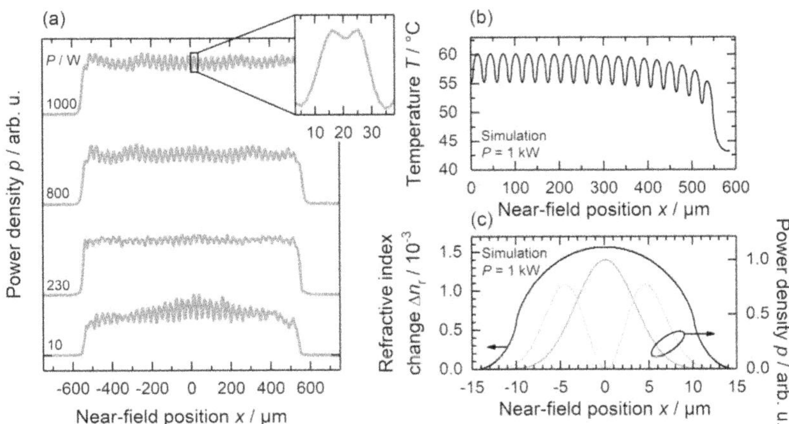

Figure 5.14: Near-field characteristic of bars with sub-structured 1095 μm-wide emitters. (a) Near-field profiles (average of the seven bar emitters) measured at high duty cycle 4% and four operation powers. (b) Simulated lateral temperature profile across a single sub-structured bar emitter. (c) Temperature-driven refractive index profile across a single sub-emitter along with the first two supported modes.

as marked in the graph. The next-order array super-mode appears at increasing operation point with peaks at $\pm 1.8°$. At 230 W the profiles exhibit little intensity around 0° indicating a high contribution from coherently coupling sub-stripes. This has changed at 800 W where the narrow super-mode features are supported by a wide Gaussian-like base. This indicates that the sub-stripes operate more and more individually as a result of the temperature (and thus refractive index) profiles building up around every sub-stripe. The coherent contributions have nearly vanished at 1 kW. Similar behavior is found for high duty cycle in Figure 5.13(c), except that the incoherent contributions are stronger at least from 800 W on. The extra shoulders appearing on both sides of the Gaussian-like base indicate higher-order modes within the individual sub-stripes.

This transition from coherent coupling at low power to incoherent emission at elevated power and heating is also visible in the respective near-field profiles. The average profile of the seven bar emitters[24] is depicted in Figure 5.14(a), measured at high duty cylce (2 ms, 20 Hz). At low power 10 W the profile is clearly modulated with its period 29 μm corresponding to the sub-emitter pitch p_{sub}. At higher power 230 W the emitters expe-

[24] All features discussed in the following are also present in the individual emitter near-field profiles and are hence not artifacts of the averaging process.

rience a transition in which the modulation depth decreases and the profiles appear less regular, as was also observed in [Spr10]. Further increase in operation power (and thus heating) lead to an again well-modulated pattern, solely with a period $p = 29\,\mu$m at 800 W and in some cases with a double-lobe feature of individual sub-emitters at 1 kW. A coarse reproduction of this double-lobe feature is done by simulating the bar temperature profile (CW, $R_{\mathrm{th}} = 0.05\,$K/W) shown in Figure 5.14(b) and calculating the modes supported by the lateral waveguide on each sub-emitter as a result of the temperature-driven refractive index change, cf. Fig. 5.14(c). The second-order mode is of a double-lobe shape with about 10 μm between its two near-field maxima. This onset of higher-order modes within the sub-stripes as observed in the near-field pattern is in line with the aforementioned appearance of wide extra shoulders in the far-field pattern.
While the just investigated lateral design of seven such sub-structured emitters processed comparably far apart is suitable to study the influence of sub-structuring in an isolated manner, better performance is achieved when bringing the emitters closer to one another and exploiting the benefit of thermal emitter cross-talk. This was shown in Section 5.2 and in particular Figure 5.7 displays the reduction in far-field width such that these sub-structured 1095 μm-wide emitters appear more suitable for high-brightness designs than apparent from the results in this section.

5.5. Conclusions

This chapter studied the beam quality of 1 kW-emitting laser bars. The determination of the governing physical mechanisms and of beneficial technological measures succeeded thanks to the emitter-resolved analysis proposed and carried out herein.

Firstly, the influence of mechanical bar deformation ("bar smile") was investigated using bars with 186 μm-wide emitters tested at low heating $R_{\mathrm{th}} \approx 0.02\,$K/W (QCW 200 μs, 10 Hz). Two bar types were contrasted, namely those featuring index-guiding trenches (baseline) and a new technology exploiting the deep-implantation technique instead. The baseline design was found sensitive to mounting-induced bar smile as this increased the far-field width to $\Theta_{95\%}(1\,\mathrm{kW}) = 11.0$, the latter rapidly varying along the bar following the strain distribution. The physical mechanism causing this degradation was then identified. In contrast to ascription in literature, both TM addition [Win14] and polarization-axis rotation [Cas13a] were found not accountable although present in the operating devices. Instead, the data suggest that photo-elastically induced waveguiding shapes and widens the TE modes which was observed as widened near-field profiles, which had continuously increasing curvature. The new design reduced both far-field width and TM-power-fraction (increased DOP) at the same smile and when mounted to yield low ($\leq 0.8\,\mu$m) smile the

far field narrowed to $\Theta_{95\%}(1\,\mathrm{kW}) = 8.8°$. This is the lowest value reported at such a high power level.
In another experiment, the effect of thermal emitter cross-talk on the lateral far-field width was investigated. A theoretical analysis of the spacing d_s between neighboring emitters at which thermal emitter cross-talk may be expected ($\forall\ d_s \leq d_{th} \approx 240\,\mu m$) led to two contrasted lateral designs using $d_{s,A} = 255\,\mu m$ and $d_{s,B} = 65\,\mu m$, respectively. Alike behavior of the two designs when plotting the far-field width as a function of emitter-power acquired at low heating for a thermal resistance $R_{th} \approx 0.02\,K/W$ (QCW 200 μs, 10 Hz) showed that thermal effects are negligible at these test conditions. Elevated heating ($R_{th} \approx 0.05\,K/W$, QCW 2 ms, 20 Hz) doubled the rate at which the far-field width increased with rising operation point only for the $d_s = 255\,\mu m$-structure. The bar design featuring close emitters $d_s = 65\,\mu m$ on the other hand revealed a rate unchanged to the low-thermal case – for all non-edge emitters – enabling substantially reduced lateral divergence. Simulations of the lateral temperature profile in the two cases showed the flattening of the T-distribution when emitters are brought closer to one another. Parabolic fits extracted the respective curvature and it increased with dissipated emitter power by half the rate for close emitters, as did the far-field width. This demonstrates the benefit of tailored thermal emitter cross-talk for low lateral divergence via reduced thermal waveguiding. The simulations further showed that the maximum temperature within the emitter was not increased with reduced spacing which from a device reliability point of view is a soothing observation.
As a direct application, bars of the lateral layout ($37 \times 186\,\mu m$) with close emitters $d_s = 65\,\mu m$ were tested at elevated heating ($R_{th} \approx 0.05\,K/W$) where they benefit from thermal emitter cross-talk. The lateral divergence at the operation point amounted to $\Theta_{95\%}(1\,\mathrm{kW}) = 10.8°$, cf. Tab. 5.3, which is the lowest value found from all tested designs. As a critical value in many applications, it is deduced which power corresponds to an 8°-divergence. The structure feeds 0.82 and 0.63 kW into this angle at the two heating scenarios, respectively, with $\eta = 63\%$ electrical-to-optical conversion efficiency in both cases.

As a direct consequence of the overall beneficial thermal emitter cross-talk, the otherwise periodic lateral temperature distribution $T(x) = T(x+w+d_s)$ observes a discontinuity at the two bar edges. In turn, the outmost emitters $i = 1,\, n_e$ are subject to a locally non-line-symmetric temperature profile, also deduced experimentally via emitter-position-resolved wavelength acquisition and supported by thermal simulations. Its effect was seen in the near-field profile which – for the respective outmost emitter – was squeezed towards the bar center. More prominent was the effect on the far field, which (i) observed a widening (1° wider than the average 8.6° of remaining emitters and 4.6° above average 9.8° at low

QCW parameters	R_{th} / KW^{-1}	P / kW	$\Theta_{1,\,95\%}$ / °	η / %
200 μs, 10 Hz	0.02	1	8.8	61
		0.82	8.0	63
2 ms, 20 Hz	0.05	1	10.8	54
		0.63	8.0	63

Table 5.3: Highest-brightness results deduced from Fig. 5.9 at (i) 1 kW emitted power and (ii) at 8° divergence, for the two heating conditions corresponding to the given thermal resistance.

and elevated heating, respectively[25]) and (ii) pointed off the forward direction by up to 1.4° with anti-symmetric behavior from one to the other bar edge. It is important to note that for the stripe widths tested herein, for which hold that $w + d_s > d_{th}$, only the respective outmost emitter was affected by the edge effect and not a gradual transition towards the bar center from emitter to emitter was observed instead. This simplifies engineering potential solutions for its mitigation. The overall result of this effect on the bar far field amounts to a widening by $[0.5 : 3]°$ for emitter widths $w \in [186 : 1095]\,\mu$m, cf. Table 5.2.

A final experiment explored the beam quality of unconventional $w = 1095\,\mu$m-wide emitters. Lateral bar designs with such stripes showed the highest efficiency found herein with 64% at 1 kW and elevated heating $R_{th} \approx 0.05$ K/W (compare 54% at same heating from $37 \times 186\,\mu$m-design) and having them, in addition, operating with narrow far field would be of great value. Such wide emitters cannot be fully contacted but require periodically structured current injection (opening $w_{sub} = 20\,\mu$m, period $p_{sub} = 29\,\mu$m herein) in order to suppress transverse lasing. This sub-structuring (here implemented via alternating implantation patterns of the p-contact layer) was found to have profound consequences on the near- and far-field profiles. Essentially, depending on the power and heating level, one observes (i) coherent coupling (low P and ΔT_{AZ}) of these sub-emitters with a phase shift of $\Delta\phi = \pi$ between neighboring sub-emitters resulting in double-peaked far-field distributions (at angles $\theta \propto \pm 1/p_{sub}$) with little emission in forward direction or (ii) individual operation of each sub-emitter due to a thermal waveguide building up around each sub-stripe guiding modes of at least second order and resulting in a bell-shaped far field with prominent shoulders. Increasing operation point P and heating ΔT_{AZ} led to a smooth transition between the two limiting cases.
It was also found that these wide emitters benefit from thermal cross-talk when brought closer together. The nature of this effect can now, that the individualized operation of the sub-emitters at high power and heating was discussed, be understood in more detail.

[25] for a $w = 186\,\mu$m, $d_s = 65\,\mu$m-bar

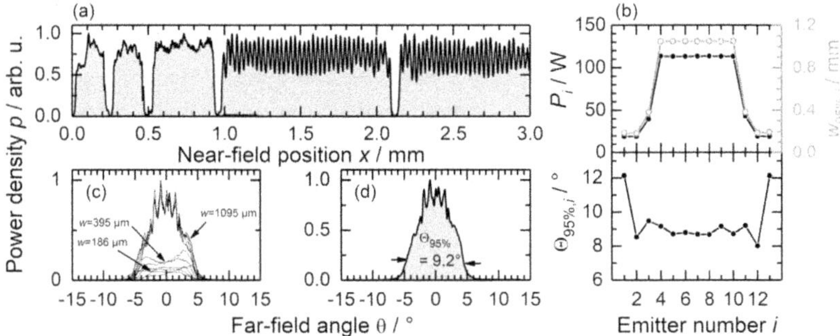

Figure 5.15: Projected performance of a lateral design with varying stripe width at elevated heating $R_{\mathrm{th}} \approx 0.05\,\mathrm{K/W}$ emitting a total power $P = 0.97\,\mathrm{kW}$ (a) Near-field profile covering the first four emitters. (b) Emitter-resolved power P_i, near-field $w_{95\%,i}$, and far-field width $\Theta_{95\%,i}$. (c) Far-field profiles of the individual emitters. (d) Overall bar far-field profile post superposition.

Sub-emitters at the emitter edge are, just like edge emitters at the bar edges, subject to a locally non-line-symmetric temperature profile and thus their emission is not only widened but also points off the forward direction leading to extra shoulders in the overall emitter far field. This was prominently seen in Figure 5.7 where profiles from the two emitter-spacing values were compared.

With respect to the achieved far-field width at the operation point 1 kW, when processed as being thermally isolated from each other at an emitter spacing $d_{\mathrm{s}} = 255\,\mu\mathrm{m}$, the resultant divergence ($\Theta_{95\%}(1\,\mathrm{kW}) = 14.5°$ at $R_{\mathrm{th}} \approx 0.05\,\mathrm{K/W}$) is not competitive. When brought closer together ($d_{\mathrm{s}} = 65\,\mu\mathrm{m}$) all inner emitters benefit from thermal cross-talk, as was already discussed, leading to a bar angle of $\approx 12.1°$. The edge emitters are responsible for $\geq 3°$ of this width.

This inspires to estimate the performance of a potential design of seven $1095\,\mu$m-wide emitters surrounded by two $186\,\mu$m-wide emitters on each bar side, all spaced at $d_{\mathrm{s}} = 65\,\mu\mathrm{m}$. To fill up the given bar width, two additional $395\,\mu$m-wide emitters were inserted. As mentioned above, the thermal edge effect was found to have decayed beyond the second emitter and thus the nine inner emitters operate with narrow far field benefiting from thermal cross-talk. The following numerical projection uses near- and far-field profiles experimentally retrieved from bars with the three stripe widths 186, 395, and $1095\,\mu$m with edge or inner emitters selected, respectively. The respective operation point was chosen for a similar injected current density $j = I/(L\sum_i w_i) \approx 25\,\mathrm{A/(mm)^2}$, which corresponds to equal applied voltage at every emitter as is the case of a biased laser bar. The com-

posed near-field profile is displayed in Figure 5.15(a) where the asymmetric edge emitter is neighbored by a line-symmetric narrow emitter and the following wider emitters. Panel (b) shows the power P_i, near-field $w_{95\%,i}$, and far-field width $\Theta_{95\%,i}$ for every emitter in this proposed lateral design. The individual far-field profiles are shown in panel (c) and the overall bar far-field profile in panel (d). At $P = 0.97\,\mathrm{kW}$ emitted power this projected design exhibits $\Theta_{95\%} = 9.2°$ lateral divergence and thus 1.4° less than the record design presented at the same power and same elevated heating.
This illustrates the high potential of continued bar development for further improved beam quality.

6. Conclusions and Outlook

This dissertation investigated GaAs-based broad-area diode-laser bars optimized for the emission of 1 kW optical power at 940 nm wavelength operating at room temperature $T_{HS} = 25°C$. Due to high industrial demand, any concept for improved bar performance, be it power scaling, efficiency increase, divergence reduction, or preferably a combination of all three, has the likely potential to be implemented by a large and dynamic market. The focus in this work was laid on the improvement of efficiency and lateral far-field width at the operation point. For this, a series of design studies was carried out with the goal of isolating the restricting physical mechanisms and to conceive technological measures in order to advance overall device performance. Each of these two design aspects yielded record results and additionally paved the way for further improvements as concrete next design steps were derived.

Design optimization for increased efficiency

The **EDAS large optical cavity** concept was found suitable for the design of the vertical (epitaxial) structure as it enabled the combination of low optical loss $\alpha_i = 0.35\,\text{cm}^{-1}$, high modal gain $\Gamma g_0 = 8.7\,\text{cm}^{-1}$, and low areal series resistance $\rho_a = R_s \cdot A \approx 9\,\text{m}\Omega\,(\text{mm})^2$. Performance extrapolation based on these device parameters projected a conversion efficiency of $\eta = 61\%$ at the operation point $P = 1$ kW, also verified in measurements of processed devices. This design of 4 mm length and a lateral layout using $n_e = 37$ emitters of $w = 186\,\mu\text{m}$ width with a spacing of $d_s = 65\,\mu\text{m}$ served as a baseline for all subsequent modifications.

Following this low-thermal testing[26], then measurements were carried out at increased duty cycle by choosing from $\Delta t_p \in [0.2 : 4]$ ms, $f_{rep} \in \{10, 20\}$ Hz. With the lasing wavelength being recorded, this allowed extracting the relation to the device heating per operation power, i.e. its thermal resistance R_{th}. The inferred non-linear increase of this dynamic "thermal impedance" Z_{th} with duty cycle was explained as the interplay of the laser's self-heating and the cooling across a thermal resistance via thermal contact to a heatsink. Modeling this dependence succeeded with a rate-equation approach and found that heating reaches down to the submount during the time spans herein assessed. Having the dependence of the thermal impedance on the QCW duty cycle explored, one now has a reproducible method at hand for **tuning the thermal resistance** between 0.01 and 0.05 K/W and so emulating the expected behavior in CW operation when advanced cooling architectures with R_{th}-values in this range are used. State-of-the-art mounting reaches values at the higher end of this span and future developments thereof will likely

[26] QCW pulse length $\Delta t_p = 200\,\mu\text{s}$, repetition rate $f_{rep} = 10$ Hz

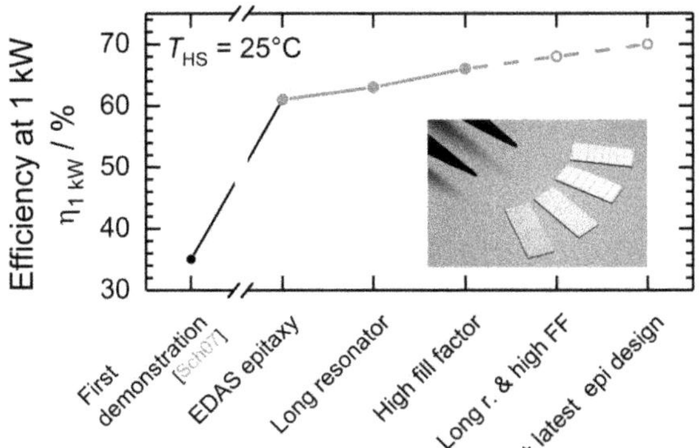

Figure 6.1: Bar-design progress for continued conversion-efficiency improvement at 1 kW. The realized designs (gray, solid) are contrasted to the first demonstration (solid, black) and projected performances (gray, open). Inset: Fully processed bar chips including baseline and new developments, ©FBH/schurian.com.

access even the lower end.

Building on the introduced vertical structure, two different design-development approaches were then pursued, both aiming to reduce the series resistance of the device by increasing its active base-plane area. For a bar with the fixed mechanical width of 1 cm this can be realized by either lengthening the longitudinal resonator or by increasing the lateral fill-factor. With this motivation bars were processed with the two **cavity lengths** 4 and 6 mm. The observed decrease in series resistance from 0.33 to 0.23 mΩ overcompensated the increase in threshold current (36 to 47 A) and the slight decrease in slope efficiency (1.10 to 1.08 W/A). The latter – crucial to the overall performance – was attributed to the low optical loss afforded by the used vertical structure. Consequently the efficiency at the operation point 1 kW was raised to 63% in the low-thermal case, which was even maintained at elevated heating up to $R_{th} \approx 0.04$ K/W (QCW 2 ms, 10 Hz). Bar designs with even longer cavity were found to require (to date unrealistically) low optical loss $\alpha_i = 0.1\ \mathrm{cm}^{-1}$ for further efficiency gains. What is more, such long-cavity designs increase production costs as fewer devices can be processed per wafer and only significant performance gains therefore make their development plausible.

Increasing the fill-factor by **lateral structuring** instead does not reduce the number of bars processible per wafer. This study investigated lateral layouts with wider emitters up to 1095 μm with emitter spacing $d_s = 65\ \mu$m enabling increased fill-factor up to 87%. The series resistance R_s was reduced correspondingly and the efficiency at the operation point continuously increased with fill-factor to a maximum of 66%. This is the highest efficiency at 1 kW ever reported from a bar. These experimental values exceed the prior extrapolation, which was based on pure series-resistance reduction, as the record designs additionally benefited from reduced roll-over the higher the fill-factor. This was particularly notable at elevated-heating testing (QCW 2 ms, 20 Hz, $R_{\text{th}} \approx 0.05$ K/W), where the slope efficiency at the operation point increased from $S(1\text{ kW}) \approx 0.55$ (baseline design) to ≈ 0.95 W/A enabling still $\eta(1\text{ kW}) = 64\%$. The observed roll-over mitigation in wide-emitter layouts is attributed to more effective extraction of the provided gain. In view of these results, even wider emitters appear appealing in order to further increase the conversion efficiency at $P = 1$ kW.

Figure 6.1 summarizes the efficiency advancements realized herein (gray, solid) and projects feasible near-term goals (gray, open). A combination of measures studied herein in concert with the use of the latest vertical structure promises $\eta(1\text{ kW}) \geq 70\%$, cf. Tab. 4.6. This is more than twice the efficiency shown by the first 1 kW-demonstrating bar, cf. Tab. 1.2.

Studies on efficient bars with narrow far field

Bar designs which benefit from the above findings for improved efficiency were then studied with respect to the lateral far field of the emission. A series of diagnostic experiments determined those physical mechanisms governing the beam quality of cm-bars. In all these investigations, the novel approach proposed and realized in this work of emitter-resolved far-field analysis proved crucial to separating, understanding, and quantifying the physical mechanisms regulating the overall bar far field.

Firstly, the influence of the mechanical deformation ("**bar smile**") post mounting was examined at low thermal conditions. The baseline design featuring index-guiding (IG) trenches was found to be sensitive to mounting-induced chip strain and delivered wide emission, with the far-field width rapidly varying along the bar following the smile profile. Despite suggestions in literature, neither TM addition nor polarization-axis rotation was found to be accountable, although present in these devices. Instead, the data suggest that photo-elastically induced waveguiding shapes and widens the TE modes, correlating with widened and continuously more curved near-field profiles. A new bar development omitted IG trenches and used the deep implantation technique to tailor the current path. At the same smile, it enabled reduced lateral divergence, which is attributed to mitigated propagation of the external tension into the critical confining layers. Low-smile mounting ($\leq 0.8\ \mu$m) then led to the far-field profile being narrowed from 10.8° (baseline) to $\Theta_{95\%}(1\text{ kW}) = 8.8°$ with the efficiency $\eta = 61\%$ maintained, cf. Fig. 6.2(a). This is the lowest reported width at such a high power.

Then turning to elevated heating at increased duty cycle 4% ($R_{\text{th}} \approx 0.05$ K/W), the

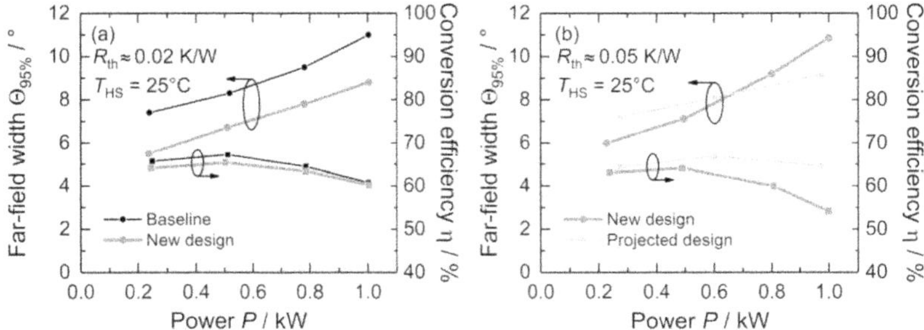

Figure 6.2: Advancement in bar design for narrow and efficient 1 kW-emission. (a) Baseline and newly developed design are contrasted at low thermal condition. (b) The novel design is compared to a projected design at elevated thermal testing.

influence of **thermal interaction between neighboring emitters** was assessed. It was found that for a spacing $d_s > d_{th} \approx 240\,\mu$m emitters operate thermally individually and that for lower values d_s the lateral temperature profiles overlap, as confirmed by thermal simulations. The resultant flattening of the temperature, and thus refractive index, distribution acted beneficially on the far-field width, which reduced thermal far-field degradation by 50%[27]. A design with $w = 186\,\mu$m, $d_s = 65\,\mu$m showed narrow emission into $\Theta_{95\%} = 10.8°$ at 1 kW and $R_{th} \approx 0.05$ K/W, while still $P = 0.63$ kW were fed into the technologically important divergence $\Theta_{95\%} = 8°$. Two important further observations were made in the simulation data: (i) emitter cross-heating did not increase the maximum temperature in the investigated emitter but solely flattened the distribution and (ii) even at the chosen small spacing $d_s = 65\,\mu$m the T-profiles still show a non-vanishing curvature at the operation point. The latter motivates future studies into further reduced emitter spacing d_s for an even flatter temperature field and thus lower divergence at the same heating. However, there may be a critical distance below which optical interaction may put proper device operation at risk.

As a direct consequence of the thermal cross-talk, the appearance of a **thermal edge effect** was observed. All emitters but those at the edge face a locally line-symmetric temperature distribution and it was shown that this perturbation only propagates by $\approx d_{th}$ from the bar edge towards its center and thus, in the case of the herein studied emitter widths $w \in [186 : 1095]\,\mu$m, affects only the outmost emitters $i = 1, n_e$ strongly. This resulted in widened far-field profiles and emission off the forward direction for these two edge emitters. The widening of the overall bar far field amounted to $[0.5 : 3]°$ for the emitter widths above, respectively. This contribution was however deduced from bar designs

[27] in the particular studied case of $w = 1095\,\mu$m contrasting $d_s = 255$ and $65\,\mu$m

with only one emitter width per bar. Its effect should shrink to well below 0.5° if the outmost emitter contributes less power to the overall bar far field. This may be realized, for example, via a reduced stripe width $w_{1,n_e} < w_{\neq 1,n_e}$, while remaining wide enough to still assure sufficient dissipated power to stabilize the next emitters $i = 2,\, n_e - 1$.
As wide emitters $w = 1095\,\mu$m proved to enable high conversion efficiency, they deserve further study with regard to their beam-quality properties. Stable lasing in the forward direction required **lateral sub-structuring** by periodic implantation of the p-contact layer forming sub-emitters (width w_{sub}, pitch p_{sub}). When tested at low-thermal conditions ($R_{th} \approx 0.02$ K/W), signatures of coherent coupling between these sub-emitters were recorded that were, as the operation point increased, gradually overshadowed by incoherent contributions. Confining temperature profiles around each sub-emitter gradually separated them and led to their individual operation. Testing at higher thermal load ($R_{th} \approx 0.05$ K/W) accelerated this transformation. Such wide emitters require thermal cross-talk in order to exhibit competitive far-field widths, and a lateral design of $n_e = 8$ emitters at $d_s = 65\,\mu$m spacing and sub-structuring ($w_{sub} = 20\,\mu$m, $p_{sub} = 29\,\mu$m) showed $\approx 12.1°$ at the operation point. Future studies may optimize the chosen sub-emitter width $w_{sub} = 20\mu$m. Initial measurements indicated that lower values ($w_{sub} = 5\mu$m, $p_{sub} = 10\mu$m tested) widen the far field, but the implementation of slightly wider sub-emitters may be beneficial.
The individualized operation of the sub-stripes at elevated heating presents another opportunity. It appears beneficial to widen the emitters even further (limiting case: entire bar width), which should (i) further increase the efficiency owing to reduced series resistance and reduced power saturation (as mentioned above), but (ii) promises to not worsen the far-field width considerably.

The results in context

The achieved brightness results are now put into perspective by comparing to the state of the art. Section 1.2 gave an overview of recent developments and summarized laser-bar results in Table 1.3. Contrasting the lateral divergence $\Theta_{95\%} = 8.8°$ measured at $R_{th} \approx 0.02$ K/W, the herein developed bar design increased the power fed into this angle from 0.27 kW [Hei18] to 1 kW, by a factor ×3.7. When comparing at similar thermal resistance $R_{th} \approx 0.05$ K/W, where herein a lateral far-field width $\Theta_{95\%} = 10.8°$ was inferred, the power fed into this angle was increased by a factor ×1.5, from 0.65 kW [Str18] to 1 kW.
Table 6.1 summarizes these 1 kW-specifications obtained in this work and computes further parameters for beam-quality quantification, which were introduced in Section 2.2. The metrics in this table are given for the low-thermal case $R_{th} \approx 0.02$K/W. For elevated-heating results, far-field width $\Theta_{95\%}$, BPP, M^2 can be derived by multiplying by a factor

$P = 1\,\mathrm{kW}$	Lateral	Vertical
$\Theta_{95\%}$ / °	8.8	48 (0.24)
$w_{95\%}$ / μm	8809 (183)$^{\#}$	1.7 (9071)
BPP / mm mrad	338 (7.0)	0.36 (9.5)
M^2	1130 (23)	1.2 (32)
B_{lin} / kW (mm mrad)$^{-1}$	0.003 (0.14)	2.8 (0.11)
B / MW (cm^2 sr)$^{-1}$	53 (94)	

Table 6.1: Beam-quality properties in lateral (measured) and vertical (simulated) direction achieved by the best bar design at QCW-test conditions (200 μs, 10 Hz) chosen for a thermal resistance $R_{\mathrm{th}} \approx 0.02$ K/W. Higher thermal resistance $R_{\mathrm{th}} \approx 0.05$ K/W (QCW, 2 ms, 20 Hz) resulted in a far-field width of 10.8° (near-field width values reduced by 2%) such that BPP and M^2 (B_{lin} and B) are increased (decreased) by a factor 1.2. $^{\#}$Values in brackets are calculated emulating the use of beam rotators, with fast-axis collimators, for near-field stacking.

1.2, and linear brightness B_{lin}, radiance B by dividing by this factor. The bar exhibits $M^2 = 1130 \times$ diffraction-limited lateral and near diffraction-limited ($M^2 = 1.2$) vertical emission, such that the overall radiance amounts to 53 MW/(cm^2 sr). For further comparison, the table also lists – given in brackets – the expected behavior when equipping the bar with beam rotators, as routinely used, e.g. in [Haa17], whereafter the lateral near-field profiles appear stacked upon each other. Notably, the resultant linear brightness observes high values $B_{\mathrm{lin}} \in [0.11 : 0.14]$ kW/(cm^2 sr) in both axes. Comparing to state-of-the-art laser systems for materials processing, cf. Tab. 1.1, the use of only a few (2...4) of such laser bars can compete with some of the direct-diode systems listed there, however, without the need for complex spectral stabilization and wavelength multiplexing commonly required.

Prospect to this work

The dynamic laser market and continued technology development will demand ever improved laser-bar performance. This work paved the way for a promising approach, where wide emitters ($w = 1095\,\mu$m) are employed, as was detailed in Figure 5.15. Emitter-resolved analysis proved particularly important in the case of these wide emitters as it allowed separating the surprisingly good performance of all inner emitters that otherwise would have been overshadowed by the strong thermal edge effect – strong, since such wide edge emitters carry a high fraction of the total bar power. The knowledge of the inner

emitters' far-field characteristic allowed projecting the $R_{\mathrm{th}} \approx 0.05$ K/W-performance of this wide emitter-based design that promises $\Theta_{95\%}(1\ \mathrm{kW}) \approx 9.2°$. This falls below the high-thermal record in this thesis, 10.8°, lowering the far-field width by further 1.6°.
In addition to improved brightness, it also promises to increase the conversion efficiency. Its estimation used[28]

$$\eta(j) = \frac{P}{(V_0 + R_{\mathrm{s}}I) \cdot I} = \frac{\sum_i P_i(j)}{(V_0 + V_{\mathrm{d}} + \rho_{\mathrm{a}} j) \cdot jL\sum_i w_i},$$

where the power P_i delivered by the individual emitters of width w_i was inferred from the respective curve in Figure 4.12(a), read off at equal current density for the three operation points $j = I/(L\sum_i w_i) = 8,\ 16,\ 25\ \mathrm{A/(mm)^2}$ and the three stripe widths w. Figure 6.2 displays the projected performance at this high heating level with $\eta(1\ \mathrm{kW}) \geq 63\%$ being inferred for this potential bar design. This is $\approx$ 10%-points above the efficiency of the highest-brightness design found herein.
This concrete development goal hence promises to further advance the state-of-the-art by establishing a good balance between maximum efficiency and narrowest far field at high power. It is among the reasons why high-power laser bars face a promising future and this thesis opened new avenues to continued performance improvement.

[28] making use of the values for V_0, V_{d}, and ρ_{a} derived in Table 4.1

Appendices

A. Beam-Path Calculation Using the Matrix Formalism

The matrix formalism for consecutive optical elements can help gaining basic insights on how an optical setup operates. It lays the paraxial approximation at its foundation and ascribes a few matrices to basic optical elements. Matrix multiplication in the order of appearance in the optical setup then forms a net matrix for the entire optical path. Two basic matrices are required in the following derivations, cf. [Hal64]. A free optical path of optical length z (length projected onto optical axis) is described with the matrix T, while a lens of focal length f and vanishing extent can be included via L.

$$T(z) = \begin{pmatrix} 1 & z \\ 0 & 1 \end{pmatrix} \qquad\qquad L(f) = \begin{pmatrix} 1 & 0 \\ -\frac{1}{f} & 1 \end{pmatrix}$$

Single Keplerian telescope

By inspection of Figure 3.5(a) the net system matrix $\mathrm{NF}_{\mathrm{single}}$ can be calculated as

$$\mathrm{NF}_{\mathrm{single}} := \mathrm{NF}(f_1, z_1', f_4) = T(f_4) \cdot L(f_4) \cdot T(f_4) \cdot T(z_1') \cdot T(f_1) \cdot L(f_1) \cdot T(f_1)$$

$$\mathrm{NF}_{\mathrm{single}} = \begin{pmatrix} -\frac{f_4}{f_1} & 0 \\ \frac{z_1'}{f_1 f_4} & -\frac{f_1}{f_4} \end{pmatrix}.$$

Doubled Keplerian telescope

Likewise, the doubled telescope (cf. Fig. 3.5(b)) can be computed from the single-telescope matrices as

$$\mathrm{NF}_{\mathrm{double}} := \mathrm{NF}(f_2, z_2, f_4) \cdot \mathrm{NF}(f_1, z_1, f_2).$$

Matrix multiplication yields

$$\mathrm{NF}_{\mathrm{double}} = \begin{pmatrix} \frac{f_4}{f_1} & 0 \\ -\frac{z_1 + z_2}{f_1 f_4} & \frac{f_1}{f_4} \end{pmatrix}.$$

Far-field configuration

Lastly, the far-field configuration's imaging matrix (cf. Fig. 3.5(c)) can be calculated via

$$\mathrm{FF} := T(f_4) \cdot L(f_4) \cdot T(z_3) \cdot \mathrm{NF}(f_1, z_1, f_2).$$

This results in

$$\mathrm{FF} = \begin{pmatrix} \frac{z_1 f_4}{f_1 f_2} & -\frac{f_1 f_4}{f_2} \\ \frac{1}{f_1}\left(\frac{f_2}{f_4} - \frac{z_1 z_3}{f_2 f_4} + \frac{z_1}{f_2}\right) & -\frac{f_1}{f_2}\left(1 - \frac{z_3}{f_4}\right) \end{pmatrix}.$$

The matrix simplifies if the herein included telescope (L1,L2) is balanced, i.e. if $z_1 = 0$.

$$\mathrm{FF}_{\text{balanced}} = \begin{pmatrix} 0 & -\frac{f_1 f_4}{f_2} \\ \frac{f_2}{f_1 f_4} & -\frac{f_1}{f_2}\left(1 - \frac{z_3}{f_4}\right) \end{pmatrix} \tag{A.1}$$

This result illustrates the nature of this angular-distribution imaging instrument. A beam $\vec{x}_1 = (x_1\ \alpha_1)$ is imaged onto position

$$x_2 = -\alpha_1 \cdot \frac{f_1 f_4}{f_2}, \tag{A.2}$$

regardless of its origin x_1 but only depending on its angle α_1.

Calibration of the far-field configuration

The following imaging matrix is derived for the purpose of calibrating the constructed system. In the corresponding experiment, the far-field configuration is used as displayed in Figure 3.5(c), except with the CCD camera being displaced from the imaging plane by a distance Δz along the optical axis. This makes the imaging matrix $\mathrm{FF}_{\Delta z}$ read

$$\mathrm{FF}_{\Delta z} = T(\Delta z) \cdot \mathrm{FF}_{\text{balanced}} = \begin{pmatrix} \Delta z \frac{f_2}{f_1 f_4} & -\frac{f_4 f_1}{f_2} - \Delta z \frac{f_1}{f_2}(1 - \frac{z_3}{f_4}) \\ \frac{f_2}{f_1 f_4} & -\frac{f_1}{f_2}\left(1 - \frac{z_3}{f_4}\right) \end{pmatrix}.$$

At this value Δz, the light source now needs to be translated from x_0 to x_1 with regard to the entire optical system and perpendicular to its axis, $\Delta x := x_1 - x_0$. This leaves the incidence angle unchanged, $\Delta\alpha = \alpha_1 - \alpha_0 = 0$, such that $\vec{\Delta x} = (\Delta x\ 0)$. With the linearity of the matrix formalism noted, one can compute the corresponding displacement in the imaging plane $\vec{\Delta x_2}$ as

$$\vec{\Delta x_2} = \mathrm{FF}_{\Delta z} \cdot \vec{\Delta x} \quad \rightarrow \quad \frac{\Delta x_2}{\Delta x} = \Delta z \ / \ \frac{f_1 f_4}{f_2}. \tag{A.3}$$

If now several values Δz are tested at Δx with the corresponding effect in the CCD plane Δx_2 recorded, the far-field imaging factor $f_{\mathrm{FF}} = f_1 f_4 / f_2$ (i.e. converting angle in DL-plane into position on CCD, cf. equation A.2) can be deduced.

B. Thermal Wavelength Dependence of Longitudinal Modes

The light in an unstabilized broad-area laser is confined between the front and rear facet forming a longitudinal Fabry-Pérot resonator [Kle01] of the length L. The resonance condition of the supported modes is given as

$$n_{\mathrm{r}}\, L = n \cdot \lambda/2 \Rightarrow \lambda = \frac{2n_{\mathrm{r}}L}{n},$$

with the mode order $n \in \mathbb{N}_{+}$. Considering the temperature dependence of GaAs' refractive index at $\lambda \approx 940\,\mathrm{nm}$ and $T \approx 300\,\mathrm{K}$

$$n_{\mathrm{r}} = 3.56\ , \quad \frac{\mathrm{d}n_{\mathrm{r}}}{\mathrm{d}T} = 3.2 \cdot 10^{-4}\,\mathrm{K}^{-1}\ \text{[Ska03]},$$

the thermal wavelength shift of a particular mode ($\mathrm{d}n/\mathrm{d}T$=0) is estimated as

$$\frac{\mathrm{d}\lambda}{\mathrm{d}T} = \frac{\lambda}{n_{\mathrm{r}}} \cdot \frac{\mathrm{d}n_{\mathrm{r}}}{\mathrm{d}T} \approx 0.084\,\mathrm{nm/K}.$$

Bibliography

[Ada83] S. Adachi and K. Oe. Internal strain and photoelastic effects in $Ga_{1-x}Al_xAs/GaAs$ and $In_{1-x}Ga_xAs_yP_{1-y}/InP$ crystals. *Journal of Applied Physics*, 54(11):6620–6627, 1983.

[Agr85] G.P. Agrawal. Lateral-mode analysis of gain-guided and index-guided semiconductor-laser arrays. *Journal of Applied Physics*, 58(8):2922–2931, 1985.

[An14] H. An, Y. Xiong, C.-L. Jiang, B. Schmidt, and G. Treusch. Methods for slow axis beam quality improvement of high power broad area diode lasers. *Proceedings of SPIE*, 8965(89650U):1–8, 2014.

[An15] H. An, C.-L. Jiang, Y. Xiong, Q. Zhang, A. Inyang, J. Felder, A. Lewin, R. Roff, S. Heinemann, B. Schmidt, and G. Treusch. Advancements in high-power high-brightness laser bars and single emitters for pumping and direct diode application. *Proceedings of SPIE*, 9348(93480I):1–6, 2015.

[Ars19] S. Arslan, G. Erbert, A. Boni, M. Wilkens, A. Maaßdorf, J. Fricke, A. Ginolas, and P. Crump. Approaches for higher power in GaAs-based broad area diode lasers. *2019 IEEE High Power Diode Lasers and Systems Conference (HPD)*, pages 51–52, 2019.

[Bac07] F. Bachmann, P. Loosen, and R. Poprawe (Eds.). *High Power Diode Lasers: Technology and Applications*. Springer, 2007.

[Bac17] A. Bachmann, C. Lauer, M. Furitsch, H. König, M. Müller, and U. Strauß. Recent brightness improvements of 976 nm high power laser bars. *Proceedings of SPIE*, 10086(1008602):1–9, 2017.

[Bai11] J.G. Bai, P. Leisher, S. Zhang, S. Elim, M. Grimshaw, C. Bai, L. Bintz, D. Dawson, L. Bao, J. Wang, M. DeVito, R. Martinsen, and J. Haden. Mitigation of thermal lensing effect as a brightness limitation of high-power broad area diode lasers. *Proceedings of SPIE*, 7953(79531F):1–7, 2011.

[Bar10] J.R. Barber. *Elasticity*. Springer Netherlands, 2010.

[Ber18] C. Bernhardt. Facet defects as a limit to high-power diode-laser bar performance: Origin, classification and damage potential. Master's thesis, Technische Hochschule Wildau, 2018.

[Bie04] M.L. Biermann, S. Duran, K. Peterson, A. Gerhardt, J.W. Tomm, A. Bercha, and W. Trzeciakowski. Spectroscopic method of strain analysis in semiconductor quantum-well devices. *Journal of Applied Physics*, 96(8):4056–4065, 2004.

[Bie07] M.L. Biermann, D.T. Cassidy, T.Q. Tien, and J.W. Tomm. Processing-induced strains at solder interfaces in extended semiconductor structures. *Journal of Applied Physics*, 101(11), 2007.

[Bot86] D. Botez and D.E. Ackley. Phase-locked arrays of semiconductor diode lasers. *IEEE Circuits and Devices Magazine*, 2(1):8–17, 1986.

[Bud00] M. Buda, G. Iordache, G.A. Acket, T.G. van der Roer, L.M.F. Kaufmann, B.H. van Roy, E. Smallbrugge, I. Moerman, and C. Sys. Stress-induced effects by the anodic oxide in ridge waveguide laser diodes. *IEEE Journal of Quantum Electronics*, 36(10):1174–1183, 2000.

[Bug16] F. Bugge, P. Crump, C. Frevert, S. Knigge, H. Wenzel, G. Erbert, and M. Weyers. Movpe growth of laser structures for high-power applications at different ambient temperatures. *Journal of Crystal Growth*, 452:258–262, 2016.

[Cas04] D.T. Cassidy, S.K.K. Lam, B. Lakshmi, and D.M. Bruce. Strain mapping by measurement of the degree of polarization of photoluminescence. *Applied Optics*, 43(9):1811–1818, 2004.

[Cas13] D.T. Cassidy. Rotation of principal axes and birefringence in III-V lasers owing to bonding strain. *Applied Optics*, 52(25):6258–6265, 2013.

[Cas13a] D.T. Cassidy, O. Rehioui, C.K. Hall, L. Béchou, Y. Deshayes, A. Kohl, T. Fillardet, and Y. Ousten. High-power diode laser bars and shear strain. *Optics Letters*, 38(10):1633–1635, 2013.

[Chu12] S.L. Chuang. *Physics of Photonic Devices.* John Wiley & Sons, 2012.

[Col93] P.D. Colbourne and D.T. Cassidy. Imaging of stresses in GaAs diode lasers using polarization-resolved photoluminescence. *IEEE Journal of Quantum Electronics*, 29(1):62–68, 1993.

[Col95] L. Coldren and S. Corzine. *Diode Lasers and Photonic Integrated Circuits.* John Wiley, 1995.

[Cru12] P. Crump, S. Böldicke, C.M. Schultz, H. Ekhteraei, H. Wenzel, and G. Erbert. Experimental and theoretical analysis of the dominant lateral waveguiding mechanism in 975 nm high power broad area diode lasers. *Semiconductor Science and Technology*, 27(4):045001, 2012.

[Cru13] P. Crump, G. Erbert, H. Wenzel, C. Frevert, C.M. Schultz, K.-H. Hasler, R. Staske, B. Sumpf, A. Maaßdorf, F. Bugge, S. Knigge, and G. Tränkle. Efficient high-power laser diodes. *IEEE Journal of Selected Topics in Quantum Electronics*,

19(4):1501211, 2013.

[Cru14] P. Crump, C. Frevert, H. Hösler, F. Bugge, S. Knigge, W. Pittroff, G. Erbert, and G. Tränkle. Cryogenic ultra-high power infrared diode laser bars. *Proceedings of SPIE*, 9002(90021I):1–11, 2014.

[Cru17] P. Crump, M.M. Karow, S. Knigge, A. Maaßdorf, G. Tränkle, J. Lotz, W. Fassbender, J. Neukum, J. Körner, R. Boedefeld, and J. Hein. Progress in joule-class diode laser bars and high brightness modules for application in long-pulse pumping of solid state amplifiers. *Proceedings of SPIE*, 10086(100860E):1–6, 2017.

[Cru21] P. Crump, A. Meissner-Schenk, T. Kaul, S. Strohmaier, M.M. Karow, A. Boni, A. Maaßdorf, D. Martin, and G. Tränkle. Increased conversion efficiency at 800 W continuous wave output from single 1-cm diode laser bars at 940 nm. *Conference on Lasers and Electro-Optics/Europe and European Quantum Electronics Conference (CLEO/Europe-EQEC 2021)*, cb-5.1, 2021.

[Cru22] P. Crump, M. Elattar, Md.J. Miah, M. Ekterai, M.M. Karow, D. Martin, A. Maaßdorf, S. McDougall, C. Holly, S. Rauch, S. Gruetzner, S. Strohmaier, and G. Tränkle. Experimental studies into the beam parameter product of GaAs high-power diode lasers. *IEEE Journal of Selected Topics in Quantum Electronics*, 28(1):1501111, 2022.

[Del17] P. Della Casa, O. Brox, J. Decker, M. Winterfeldt, P. Crump, H. Wenzel, and M. Weyers. High-power broad-area buried-mesa lasers. *Semiconductor Science and Technology*, 32(6):065009, 2017.

[Die00] R. Diehl, editor. *High-Power Diode Lasers: Fundamentals, Technology, Applications*. Springer, 2000.

[Eic04] J. Eichler, L. Dünkel, and B. Eppich. Die Strahlgüte von Lasern - Wie bestimmt man Beugungsmaßzahl und Strahldurchmesser in der Praxis ? *Laser Technik Journal*, 2:63–66, 2004.

[Epp13] P.W. Epperlein. *Semiconductor Laser Engineering, Reliability and Diagnostics: A Practical Approach to High Power and Single Mode Devices*. John Wiley & Sons, 2013.

[Epp18] B. Eppich and G. Mann. BeamXpertDESIGNER - Software, www.beamxpert.com, 2018.

[Erb00] G. Erbert, A. Bärwolff, J. Sebastian, and J. Tomm. High-power broad-area diode lasers and laser bars. In *High-Power Diode Lasers: Fundamentals, Technology, Applications*, pages 173–223. Springer, 2000.

[Fou22] J. Fourier. *Théorie analytique de la chaleur.* Firmin Didot, père et fils, 1822.

[Fre15] C. Frevert, P. Crump, F. Bugge, S. Knigge, A. Ginolas, and G. Erbert. Low-temperature optimized 940 nm diode laser bars with 1.98 kW peak power at 203 K. *2015 Conference on Lasers and Electro-Optics (CLEO)*, SM3F.8, 2015.

[Fre16] C. Frevert, F. Bugge, S. Knigge, A. Ginolas, G. Erbert, and P. Crump. 940 nm QCW diode laser bars with 70% efficiency at 1 kW output power at 203 K: analysis of remaining limits and path to higher efficiency and power at 200 K and 300 K. *Proceedings of SPIE*, 9733(9733L):1–13, 2016.

[Fre19] C. Frevert. *Optimization of broad-area GaAs diode lasers for high powers and high efficiencies in the temperature range 200-220 K.* PhD thesis, Technische Universität Berlin, 2019.

[Gef20] B. Gefvert, C. Holton, A. Nogee, and J. Hecht. Annual laser market review & forecast 2020: Laser markets navigate turbulent times. *Laser Focus World*, January 2020.

[Haa17] M. Haas, S. Rauch, S. Nagel, R. Beißwanger, T. Dekorsy, and H. Zimer. Beam quality deterioration in dense wavelength beam-combined broad-area diode lasers. *IEEE Journal of Quantum Electronics*, 53(3):1–11, 2017.

[Hal64] K. Halbach. Matrix representation of gaussian optics. *American Journal of Physics*, 32(2):90–108, 1964.

[Has14] K.H. Hasler, H. Wenzel, P. Crump, S. Knigge, A. Maaßdorf, R. Platz, R. Staske, and G. Erbert. Comparative theoretical and experimental studies of two designs of high-power diode lasers. *Semiconductor Science and Technology*, 29(4):045010, 2014.

[Hec09] E. Hecht. *Optik.* Oldenbourg, 5th edition, 2009.

[Hei18] S. Heinemann, S.D. McDougall, G. Ryu, L. Zhao, X. Liu, C. Holly, C.-L. Jiang, P. Modak, Y. Xiong, T. Vethake, S.G. Strohmaier, B. Schmidt, and H. Zimer. Advanced chip designs and novel cooling techniques for brightness scaling of industrial, high power diode laser bars. *Proceedings of SPIE*, 10514(105140Y):1–9, 2018.

[Hem12] M. Hempel, M. Ziegler, S. Schwirzke-Schaaf, J.W. Tomm, D. Jankowski, and D. Schröder. Spectroscopic analysis of packaging concepts for high-power diode laser bars. *Applied Physics A*, 107(2):371–377, 2012.

[Hig69] C.W. Higginbotham, M. Cardona, and F.H. Pollak. Intrinsic piezobirefringence of Ge, Si, and GaAs. *Physical Review*, 184:821–829, 1969.

[Hol18] C. Holly. *Modeling of the lateral emission characteristics of high-power edge-emitting semiconductor lasers.* PhD thesis, Rheinisch-Westfälische Technische Hochschule Aachen, 2018.

[Hol18a] C. Holton, G. Overton, A. Nogee, and K. Kincade. Annual laser market review & forecast: Lasers enabling lasers. *Laser Focus World*, January 2018.

[Hoo78] R. Hook. *Lectures de Potentia Restitutiva, Or of Spring.* John Martyn, 1678.

[Joy82] W. Joyce. Role of the conductivity of the confining layers in DH-laser spatial hole burning effects. *IEEE Journal of Quantum Electronics*, 18(12):2005–2009, 1982.

[Kan18] M. Kanskar, S. Keeney, and R. Martinsen. The power of brilliance - the past and future of high-power semiconductor lasers. *Laser Focus World*, January 2018.

[Kar17] M.M. Karow, T. Kaul, S. Knigge, A. Maaßdorf, G. Erbert, S.G. Strohmaier, and P. Crump. Long-resonator laser-diode bars for efficient kW emission. *Conference on Lasers and Electro-Optics/Europe and European Quantum Electronics Conference (CLEO/Europe-EQEC 2017)*, cb-7.3, 2017.

[Kar17a] M.M. Karow, C. Frevert, R. Platz, S. Knigge, A. Maaßdorf, G. Erbert, and P. Crump. Efficient 600-W-laser bars for long-pulse pump applications at 940 and 975 nm. *IEEE Photonics Technology Letters*, 29(19):1683–1686, 2017.

[Kar19] M.M. Karow, D. Martin, P. Della Casa, G. Erbert, and P. Crump. Narrower far field and higher efficiency in 1 kW diode-laser bars using improved lateral structuring. *Conference on Lasers and Electro-Optics/Europe and European Quantum Electronics Conference (CLEO/Europe-EQEC 2019)*, cb-5.4, 2019.

[Kar20] M.M. Karow, D. Martin, P. Della Casa, G. Erbert, and P. Crump. Design progress for higher efficiency and brightness in 1 kW diode-laser bars. *Proceedings of SPIE*, 11262(1126205):1–7, 2020.

[Kau19] T. Kaul, G. Erbert, A. Klehr, A. Maaßdorf, D. Martin, and P. Crump. Impact of carrier nonpinning effect on thermal power saturation in GaAs-based high power diode lasers. *IEEE Journal of Selected Topics in Quantum Electronics*, 25(6):1–10, 2019.

[Kir79] P.A. Kirkby, P.R. Selway, and L.D. Westbrook. Photoelastic waveguides and their effect on stripe-geometry GaAs/$Ga_{1-x}Al_xAs$ lasers. *Journal of Applied Physics*, 50(7):4567–4579, 1979.

[Kit15] M. Kitzmantel and E. Neubauer. Innovative hybrid heat sink materials with high thermal conductivities and tailored CTE. *Proceedings of SPIE*, 9346(934606):1–11, 2015.

[Kle01] A. Klehr, G. Beister, G. Erbert, A. Klein, J. Maege, I. Rechenberg, J. Sebastian, H. Wenzel, and G. Tränkle. Defect recognition via longitudinal mode analysis of high power fundamental mode and broad area edge emitting laser diodes. *Journal of Applied Physics*, 90(1):43–47, 2001.

[Kna11] M.T. Knapczyk, J.H. Jacob, H. Eppich, A.K. Chin, K.D. Lang, J.T. Vignati, and R.H. Chin. 70% efficient near 1kW single 1-cm laser-diode bar at 20°C. *Proceedings of SPIE*, 7918(79180F):1–6, 2011.

[Laß00] K. Laßwitz. *Wirklichkeit – Beiträge zum Weltverständnis.* Verlag von Emil Feber, 1900.

[Li08] H. Li, F. Reinhardt, I. Chyr, X. Jin, K. Kuppuswamy, T. Towe, D. Brown, O. Romero, D. Liu, R. Miller, T. Nguyen, T. Crum, T. Truchan, E. Wolak, J. Mott, and J. Harrison. High-efficiency, high-power diode laser chips, bars, and stacks. *Proceedings of SPIE*, 6876(68760G):1–6, 2008.

[Liu17] X. Liu, C. Holly, C. Jiang, Y. Xiong, S. McDougall, K. Boucke, H. Zimer, and B. Schmidt. Electro-optical efficiency and slow axis far-field improvement of high power laser diode bars using epitaxy structure optimization. *2017 IEEE High Power Diode Lasers and Systems Conference (HPD)*, 2017.

[Mar19] D. Martin, P. Della Casa, T. Adam, C. Goerke, A. Thies, K. Häusler, O. Brox, H. Wenzel, P. Crump, M. Weyers, and A. Knigge. Current spreading suppression by O- and Si-implantation in high power broad area diode lasers. *Proceedings of SPIE*, 10900(109000M):1–7, 2019.

[May12] H.C. Mayer and R. Krechetnikov. Walking with coffee: Why does it spill? *Physical Review E*, 85:046117, 2012.

[McD20] S.D. McDougall, T. Barnowski, G. Ryu, S. Heinemann, T.Vethake, X.Liu, C.-L. Jiang, and H. Zimer. Advances in diode laser bar power and reliability for multi-kW disk laser pump sources. *Proceedings of SPIE*, 11262(1126206):1–9, 2020.

[Nye85] J.F. Nye. *Physical Properties of Crystals: Their Representation by Tensors and Matrices.* Clarendon Press, 1985.

[Pea90] S.J. Pearton. Ion implantation for isolation of III-V semiconductors. *Materials Science Reports*, 4(6):313–363, 1990.

[Pip13] J. Piprek. *Semiconductor Optoelectronic Devices: Introduction to Physics and Simulation.* Academic Press, 2013.

[Pip13a] J. Piprek and Z.M. Simon Li. On the importance of non-thermal far-field blooming in broad-area high-power laser diodes. *Applied Physics Letters*, 102(22):221110, 2013.

[Pit01] W. Pittroff, G. Erbert, G. Beister, F. Bugge, A. Klein, A. Knauer, J. Mäge, P. Ressel, J. Sebastian, R. Staske, and G. Tränkle. Mounting of high power laser diodes on boron nitride heat sinks using an optimized Au/Sn metallurgy. *IEEE Transactions on Advanced Packaging*, 24(4):434–441, 2001.

[Pla14] R. Platz, B. Eppich, P. Crump, W. Pittroff, S. Knigge, A. Maaßdorf, and G. Erbert. 940-nm broad area diode lasers optimized for high pulse-power fiber coupled applications. *IEEE Photonics Technology Letters*, 26(6):625–628, 2014.

[Pod19] A.A. Podoskin, D.N. Romanovich, I.S. Shashkin, P.S. Gavrina, Z.N. Sokolova, S.O. Slipchenko, and N.A. Pikhtin. Specific features of closed-mode formation in rectangular resonators based on InGaAs/AlGaAs/GaAs heterostructures for high-power semiconductor lasers. *Physics of Semiconductor Devices*, 53(6):828–832, 2019.

[Rau17] S. Rauch, H. Wenzel, M. Radziunas, M. Haas, G. Tränkle, and H. Zimer. Impact of longitudinal refractive index change on the near-field width of high-power broad-area diode lasers. *Applied Physics Letters*, 110(26):263504, 2017.

[Ray79] Lord Rayleigh. Investigations in optics, with special reference to the spectroscope. *The London, Edinburgh, and Dublin Philosophical Magazine and Journal of Science: Series 5*, 8(49):261–274, 1879.

[Res05] P. Ressel, G. Erbert, U. Zeimer, K. Häusler, G. Beister, B. Sumpf, A. Klehr, and G. Tränkle. Novel passivation process for the mirror facets of Al-free active-region high-power semiconductor diode lasers. *IEEE Photonics Technology Letters*, 17(5):962–964, 2005.

[Ried18] S. Ried, S. Rauch, L. Irmler, J. Rikels, A. Killi, E. Papastathopoulos, E. Sarailou, and H. Zimer. Next generation diode lasers with enhanced brightness. *Proceedings of SPIE*, 10514(105140G):1–8, 2018.

[Rie18] J. Rieprich, M. Winterfeldt, R. Kernke, J. W. Tomm, and P. Crump. Chip-carrier thermal barrier and its impact on lateral thermal lens profile and beam parameter product in high power broad area lasers. *Journal of Applied Physics*, 123(12):125703, 2018.

[Rie19] J. Rieprich, G. Blume, A. Ginolas, S. Arslan, R. Kernke, J.W. Tomm, K. Paschke, and P. Crump. Thermal boundary resistance between GaAs and p-side metal as limit to high power diode lasers. *2019 IEEE High Power Diode Lasers and Systems Conference (HPD)*, pages 35–36, 2019.

[Sch07] D. Schröder, J. Meusel, P. Hennig, D. Lorenzen, M. Schröder, R. Hülsewede, and J. Sebastian. Increased power of broad-area lasers (808 nm/980 nm) and applicability to 10-mm bars with up to 1000 Watt QCW. *Proceedings of SPIE*, 6456(64560N):1–10, 2007.

[Sch20] G. Schafferus. Analyse von Facettendefekten an roten Hochleistungs-Diodenlaser-Barren. Master's thesis, Beuth Hochschule für Technik Berlin, 2020.

[Sie98] A.E. Siegman. How to (maybe) measure laser beam quality. In *DPSS (Diode Pumped Solid State) Lasers: Applications and Issues*, 1998.

[Ska03] T. Skauli, P.S. Kuo, K.L. Vodopyanov, T.J. Pinguet, O. Levi, L.A. Eyres, J.S. Harris, M.M. Fejer, B. Gerard, L. Becouarn, and E. Lallier. Improved dispersion relations for GaAs and applications to nonlinear optics. *Journal of Applied Physics*, 94(10):6447–6455, 2003.

[Smi76] C.J. Smithells. *Metals Reference Book*. Butterworth-Heinemann, 1976.

[Spr10] M. Spreemann, B. Eppich, F. Schnieder, H. Wenzel, and G. Erbert. Modal behavior, spatial coherence, and beam quality of a high-power gain-guided laser array. *IEEE Journal of Quantum Electronics*, 46(11):1619–1625, 2010.

[Str17] S.G. Strohmaier, G. Erbert, A.H. Meissner-Schenk, M. Lommel, B. Schmidt, T. Kaul, M. Karow, and P. Crump. kW-class diode laser bars. *Proceedings of SPIE*, 10086(100860C), 2017.

[Str18] S.G. Strohmaier, G. Erbert, T. Rataj, A.H. Meissner-Schenk, V. Loyo-Maldonado, C. Carstens, H. Zimer, B. Schmidt, T. Kaul, M.M. Karow, M. Wilkens, and P. Crump. Forward development of kW-class power diode laser bars. *Proceedings of SPIE*, 10514(1051409):1–6, 2018.

[Sun13] W. Sun, R. Pathak, G. Campbell, H. Eppich, J.H. Jacob, A. Chin, and J. Fryer. Higher brightness laser diodes with smaller slow axis divergence. *Proceedings of SPIE*, 8605(86050D):1–9, 2013.

[Sze07] S.M. Sze and K.N. Kwok. *Physics of Semiconductor Devices*. John Wiley & Sons, 2007.

[Tom06] J.W. Tomm, T.Q. Tien, and D.T. Cassidy. Spectroscopic strain measurement methodology: Degree-of-polarization photoluminescence versus photocurrent spectroscopy. *Applied Physics Letters*, 88(13):133504, 2006.

[Var67] Y.P. Varshni. Temperature dependence of the energy gap in semiconductors. *Physica*, 34(1):149–154, 1967.

[Wen90] H. Wenzel and H.-J. Wünsche. A model for the calculation of the threshold current of SCH-MQW-SAS lasers. *physica status solidi (a)*, 120(2):661–673, 1990.

[Wen13] H. Wenzel. Basic aspects of high-power semiconductor laser simulation. *IEEE Journal of Selected Topics in Quantum Electronics*, 19(5):1–13, 2013.

[Wen13a] H. Wenzel, P. Crump, J. Fricke, P. Ressel, and G. Erbert. Suppression of higher-order lateral modes in broad-area diode lasers by resonant anti-guiding. *IEEE Journal of Quantum Electronics*, 49(12):1102–1108, 2013.

[Wes08] T. Westphalen, M. Leers, C. Scholz, and K. Boucke. Emitter resolved analysis of packaged laser bars. *Proceedings of SPIE*, 6876(68760O):1–10, 2008.

[Wes13] C. Wessling, O. Rübenach, S. Hambücker, V. Sinhoff, S. Banerjeea, K. Ertel, and P. Mason. Efficient pumping of inertial fusion energy lasers. *Proceedings of SPIE*, 8602(860217):1–10, 2013.

[Win14] M. Winterfeldt, P. Crump, H. Wenzel, G. Erbert, and G. Tränkle. Experimental investigation of factors limiting slow axis beam quality in 9xx nm high power broad area diode lasers. *Journal of Applied Physics*, 116(6):063103, 2014.

[Win15] M. Winterfeldt, P. Crump, S. Knigge, A. Maaßdorf, U. Zeimer, and G. Erbert. High beam quality in broad area lasers via suppression of lateral carrier accumulation. *IEEE Photonics Technology Letters*, 27(17):1809–1812, 2015.

[Win16] M. Winterfeldt, J. Rieprich, S. Knigge, A. Maaßdorf, M. Hempel, R. Kernke, J.W. Tomm, G.Erbert, and P. Crump. Assessing the influence of the vertical epitaxial layer design on the lateral beam quality of high-power broad area diode lasers. *Proceedings of SPIE*, 9733(97330O):1–9, 2016.

[Win18] M. Winterfeldt. *Investigation of slow-axis beam quality degradation in high-power broad area diode lasers*. PhD thesis, Technische Universität Berlin, 2018.

[Yan95] J. Yang and D.T. Cassidy. Strain measurement and estimation of photoelastic effects and strain-induced optical gain change in ridge waveguide lasers. *Journal of Applied Physics*, 77(7):3382–3387, 1995.

[Zeg19] A. Zeghuzi, M. Radziunas, H. Wünsche, J.-P. Koester, H. Wenzel, U. Bandelow, and A. Knigge. Traveling wave analysis of non-thermal far-field blooming in high-power broad-area lasers. *IEEE Journal of Quantum Electronics*, 55(2):1–7, 2019.

[Zie17] J.F. Ziegler. SRIM - The stopping and range of ions in matter - Software, www.srim.org, accessed Sept. 2017.

Public Contributions

Based on the research conducted in the course of this thesis, the author has made the following public contributions.

Authorship:

- M.M. Karow, D. Martin, P. Della Casa, G. Erbert, and P. Crump, "Design progress for higher efficiency and brightness in 1 kW diode-laser bars," *Proc. SPIE*, 11262(1126205), 2020.
- M.M. Karow, D. Martin, P. Della Casa, G. Erbert, and P. Crump, "Narrower far field and higher efficiency in 1 kW diode-laser bars using improved lateral structuring," *CLEO/Europe-EQEC 2019*, cb-5.4, 2019.
- M.M. Karow, C. Frevert, R. Platz, S. Knigge, A. Maaßdorf, G. Erbert, and P. Crump, "Efficient 600-W-laser bars for long-pulse pump applications at 940 and 975 nm," *IEEE Photon. Technol. Lett.*, 29(19):1683-1686, 2017.
- M.M. Karow, T. Kaul, S. Knigge, A. Maaßdorf, G. Erbert, S.G. Strohmaier, and P. Crump, "Long-resonator laser-diode bars for efficient kW emission," *CLEO/Europe-EQEC 2017*, cb-7.3, 2017.

Co-authorship:

- P. Crump, M. Elattar, Md. J. Miah, M. Ekterai, M.M. Karow, D. Martin, A. Maaßdorf, S. McDougall, C. Holly, S. Rauch *et al.*, "Experimental studies into the beam parameter product of GaAs high-power diode lasers," *IEEE J. Sel. Top. Quantum Electron.*, 28(1):1501111, 2022.
- P. Crump, M. Elattar, Md. J. Miah, M. Ekterai, M.M. Karow, D. Martin, P. Della Casa, A. Maaßdorf, S. McDougall, C. Holly *et al.*, "Progress in experimental studies into the beam parameter product of GaAs-based high-power diode lasers," *Proc. SPIE*, 11983(1198307), 2022.
- P. Crump, A. Meissner-Schenk, T. Kaul, S. Strohmaier, M.M. Karow, A. Boni, A. Maaßdorf, D. Martin, and G. Tränkle, "Increased conversion efficiency at 800 W continuous wave output from single 1-cm diode laser bars at 940 nm," *CLEO/Europe-EQEC 2021*, cb-5.1, 2021.
- S.G. Strohmaier, G. Erbert, T. Rataj, A.H. Meissner-Schenk, V. Loyo-Maldonado, C. Carstens, H. Zimer, B. Schmidt, T. Kaul, M.M. Karow *et al.*, "Forward development

of kW-class power diode laser bars," *Proc. SPIE*, 10514(1051409), 2018.

- P. Crump, M.M. Karow, S. Knigge, A. Maaßdorf, G. Tränkle, J. Lotz, W. Fassbender, J. Neukum, J. Körner, R. Boedefeld *et al.*, "Progress in joule-class diode laser bars and high brightness modules for application in long-pulse pumping of solid state amplifiers," *Proc. SPIE*, 10086(10086E), 2017.
- J. Körner, R. Boedefeld, J. Hein, M.M. Karow, C. Frevert, P. Crump, and T. Töpfer, "The next generation of high radiance laser diode modules," *10th HEC-DPSSL*, 2017.

Patent applications:

- Matthias M. Karow, Paul Crump, and Jarez Miah "Laserbarren mit verringerter lateraler Fernfelddivergenz" PCT/EP2022/065024
- Jarez Miah, Matthias M. Karow, and Paul Crump "Laserbarren mit verringerter lateraler Fernfelddivergenz" DE 10 2021 114 411.6
- Pietro Della Casa, Paul Crump, Mohamed Elattar, and Matthias M. Karow, "Laserdiode mit integrierter thermischer Blende" PCT/EP2021/085730
- Pietro Della Casa, Paul Crump, Mohamed Elattar, and Matthias M. Karow, "Laserdiode mit integrierter thermischer Blende" DE 10 2020 133 368.4

Danksagung

Zu ganz besonderem Dank verpflichtet bin ich meinem Doktorvater, Herrn Professor Dr. Günther Tränkle, Wissenschaftlicher Direktor des Ferdinand-Braun-Instituts, Leibniz-Institut für Höchstfrequenztechnik, aufgrund seiner ausgeprägten und nachhaltigen Unterstützung wissenschaftlicher und organisatorischer Art, durch die diese Dissertation möglich wurde.

Verbunden bin ich auch den Herren Professor Dr. Michael Kneissl, Institut für Festkörperphysik, Technische Universität Berlin und Professor Stephen Sweeney, PhD, Department of Physics, University of Surrey für die Übernahme der Gutachtertätigkeit.

Nachdrücklich danke ich den Mitarbeitern des Ferdinand-Braun-Instituts, so etwa Paul Crump, PhD als enthusiastischer Leiter des Hochleistungslaser-Laboratoriums, Dr. Bernd Eppich als stets hilfreicher und erfrischender Experte der Simulation und Realisierung optischer Systeme, Dipl.-Phys. Arnim Ginolas und Sabrina Kreutzmann für die professionelle Montage vieler Barren-Chips und allen, die ebenfalls in vielfältiger Weise zum Gelingen dieser Arbeit beigetragen haben.

Ganz besonders wichtig beim Absolvieren eines solchen Langstreckenlaufes sind liebe Kollegen und Freunde, wobei ich insbesondere Jan-Philipp Koester, M.Sc., Dr. Martin Winterfeldt, Dr. Anissa Zeghuzi und Dr. Carlo Frevert für viele schöne Momente dankbar bin.

Die TRUMPF Laser GmbH, mit dem Leiter der Niederlassung Berlin, Dr. Stephan Strohmaier, hat diese Arbeit essentiell finanziell und organisatorisch unterstützt. Mein Dank geht auch an Dipl.-Ing. Stefan Grützner für die Zurverfügungstellung einzelner thermischer Simulationen.

Besten Dank an Pierce Munnelly, M.Sc. für die Hilfestellung beim sprachlichen Feinschliff.

Ohne einen sicheren Anlaufpunkt, wie ihn mir meine Familie, mein Freundeskreis und im Besonderen meine Partnerin Nora bietet, wären viele Schritte meines Lebens nicht möglich gewesen – mein tiefster Dank dafür.

Die Bedeutung von [May12] bei der Erarbeitung einer Dissertation darf ebenfalls nicht unterschätzt werden.

Biographical Sketch

Matthias Karow graduated from 1524-founded Gymnasium Ernestinum, Gotha, Germany and earned his Bachelors degree in Physics from Georg August University of Göttingen in 2011. Under the Fulbright Scholarship, he then visited Arizona State University, Tempe, USA where he joined the NSF-funded Quantum Energy and Sustainable Solar Technologies ERC (QESST) conducting research on epitaxy and characterization of multiple-quantum-well solar-cell structures. Interposed work at OSRAM Opto Semiconductors, Regensburg in 2013 familiarized him with opto-electronic applications of (AlInGa)N compounds, with the detection of hot carriers from Auger-recombination-mediated efficiency losses in LED structures as a particularly enjoyed endeavor.

After earning his Masters degree in Electrical Engineering from Arizona State University in 2014, he joined the Quantum Device Group at the Institute of Solid State Physics, Technical University of Berlin, from where he received a Masters degree in Physics in 2016. His work dealt with the characterization and modeling of integrated nano-photonic devices. Since 2015 he has been with the Ferdinand-Braun-Institut, Leibniz-Institut für Höchstfrequenztechnik, Berlin, engaging in the development of kW-class diode-laser bars and research on their performance-limiting causes, pursued in close collaboration with TRUMPF Laser GmbH.

Mr. Karow is a member of the German Physical Society (DPG), the Institute of Electrical and Electronics Engineering (IEEE), and the HKN Engineering Honor Society.

Innovationen mit Mikrowellen und Licht
Forschungsberichte aus dem Ferdinand-Braun-Institut, Leibniz-Institut für Höchstfrequenztechnik

Herausgeber: Prof. Dr. G. Tränkle

Band 1: **Thorsten Tischler**
Die Perfectly-Matched-Layer-Randbedingung in der Finite-Differenzen-Methode im Frequenzbereich: Implementierung und Einsatzbereiche
ISBN: 3-86537-113-2, 19,00 EUR, 144 Seiten

Band 2: **Friedrich Lenk**
Monolithische GaAs FET- und HBT-Oszillatoren mit verbesserter Transistormodellierung
ISBN: 3-86537-107-8, 19,00 EUR, 140 Seiten

Band 3: **R. Doerner, M. Rudolph (eds.)**
Selected Topics on Microwave Measurements, Noise in Devices and Circuits, and Transistor Modeling
ISBN: 3-86537-328-3, 19,00 EUR, 130 Seiten

Band 4: **Matthias Schott**
Methoden zur Phasenrauschverbesserung von monolithischen Millimeterwellen-Oszillatoren
ISBN: 978-3-86727-774-0, 19,00 EUR, 134 Seiten

Band 5: **Katrin Paschke**
Hochleistungsdiodenlaser hoher spektraler Strahldichte mit geneigtem Bragg-Gitter als Modenfilter (α-DFB-Laser)
ISBN: 978-3-86727-775-7, 19,00 EUR, 128 Seiten

Band 6: **Andre Maaßdorf**
Entwicklung von GaAs-basierten Heterostruktur-Bipolartransistoren (HBTs) für Mikrowellenleistungszellen
ISBN: 978-3-86727-743-3, 23,00 EUR, 154 Seiten

Band 7: **Prodyut Kumar Talukder**
Finite-Difference-Frequency-Domain Simulation of Electrically Large Microwave Structures using PML and Internal Ports
ISBN: 978-3-86955-067-1, 19,00 EUR, 138 Seiten

Band 8: **Ibrahim Khalil**
Intermodulation Distortion in GaN HEMT
ISBN: 978-3-86955-188-3, 23,00 EUR, 158 Seiten

Band 9: **Martin Maiwald**
Halbleiterlaser basierte Mikrosystemlichtquellen für die Raman-Spektroskopie
ISBN: 978-3-86955-184-5, 19,00 EUR, 134 Seiten

Band 10: **Jens Flucke**
Mikrowellen-Schaltverstärker in GaN- und GaAs-Technologie Designgrundlagen und Komponenten
ISBN: 978-3-86955-304-7, 21,00 EUR, 122 Seiten

Cuvillier Verlag
Internationaler wissenschaftlicher Fachverlag

Innovationen mit Mikrowellen und Licht
Forschungsberichte aus dem Ferdinand-Braun-Institut, Leibniz-Institut für Höchstfrequenztechnik

Herausgeber: Prof. Dr. G. Tränkle

Band 11: **Harald Klockenhoff**
Optimiertes Design von Mikrowellen-Leistungstransistoren und Verstärkern im X-Band
ISBN: 978-3-86955-391-7, 26,75 EUR, 130 Seiten

Band 12: **Reza Pazirandeh**
Monolithische GaAs FET- und HBT-Oszillatoren mit verbesserter Transistormodellierung
ISBN: 978-3-86955-107-8, 19,00 EUR, 140 Seiten

Band 13: **Tomas Krämer**
High-Speed InP Heterojunction Bipolar Transistors and Integrated Circuits in Transferred Substrate Technology
ISBN: 978-3-86955-393-1, 21,70 EUR, 140 Seiten

Band 14: **Phuong Thanh Nguyen**
Investigation of spectral characteristics of solitary diode lasers with integrated grating resonator
ISBN: 978-3-86955-651-2, 24,00 EUR, 156 Seiten

Band 15: **Sina Riecke**
Flexible Generation of Picosecond Laser Pulses in the Infrared and Green Spectral Range by Gain-Switching of Semiconductor Lasers
ISBN: 978-3-86955-652-9, 22,60 EUR, 136 Seiten

Band 16: **Christian Hennig**
Hydrid-Gasphasenepitaxie von versetzungsarmen und freistehenden GaN-Schichten
ISBN: 978-3-86955-822-6, 27,00 EUR, 162 Seiten

Band 17: **Tim Wernicke**
Wachstum von nicht- und semipolaren InAlGaN-Heterostrukturen für hocheffiziente Licht-Emitter
ISBN: 978-3-86955-881-3, 23,40 EUR, 138 Seiten

Band 18: **Andreas Wentzel**
Klasse-S Mikrowellen-Leistungsverstärker mit GaN-Transistoren
ISBN: 978-3-86955-897-4, 29,65 EUR, 172 Seiten

Band 19: **Veit Hoffmann**
MOVPE growth and characterization of (In,Ga)N quantum structures for laser diodes emitting at 440 nm
ISBN: 978-3-86955-989-6, 18,00 EUR, 118 Seiten

Band 20: **Ahmad Ibrahim Bawamia**
Improvement of the beam quality of high-power broad area semiconductor diode lasers by means of an external resonator
ISBN: 978-3-95404-065-0, 21,00 EUR, 126 Seiten

Cuvillier Verlag
Internationaler wissenschaftlicher Fachverlag

Innovationen mit Mikrowellen und Licht
Forschungsberichte aus dem Ferdinand-Braun-Institut, Leibniz-Institut für Höchstfrequenztechnik

Herausgeber: Prof. Dr. G. Tränkle

Band 21: **Agnietzka Pietrzak**
Realization of High Power Diode Lasers with Extremely Narrow Vertical Divergence
ISBN: 978-3-95404-066-7, 27,40 EUR, 144 Seiten

Band 22: **Eldad Bahat-Treidel**
GaN-based HEMTs for High Voltage Operation
Design, Technology and Characterization
ISBN: 978-3-95404-094-0, 41,10 EUR, 220 Seiten

Band 23: **Ponky Ivo**
AlGaN/GaN HEMTs Reliability:
Degradation Modes and Anslysis
ISBN: 978-3-95404-259-3, 23,55 EUR, 132 Seiten

Band 24: **Stefan Spießberger**
Compact Semiconductor-Based Laser Sources
with Narrow Linewidth and High Output Power
ISBN: 978-3-95404-261-6, 24,15 EUR, 140 Seiten

Band 25: **Silvio Kühn**
Mikrowellenoszillatoren für die Erzeugung von atmosphärischen Mikroplasmen
ISBN: 978-3-95404-378-1, 21,85 EUR, 112 Seiten

Band 26: **Sven Schwertfeger**
Experimentelle Untersuchung der Modensynchronisation in Multisegment-Laserdioden zur Erzeugung kurzer optischer Pulse bei einer Wellenlänge von 920 nm
ISBN: 978-3-95404-471-9, 29,45 EUR, 150 Seiten

Band 27: **Christoph Matthias Schultz**
Analysis and mitigation of the factors limiting the effiency of high power distributed feedback diode lasers
ISBN: 978-3-95404-521-1, 68,40 EUR, 388 Seiten

Band 28: **Luca Redaelli**
Design and fabrication of GaN-based laser diodes for single-mode and narrow-linewidth applications
ISBN: 978-3-95404-586-0, 29,70 EUR, 176 Seiten

Band 29: **Martin Spreemann**
Resonatorkonzepte für Hochleistungs-Diodenlaser
mit ausgedehnten lateralen Dimensionen
ISBN: 978-3-95404-628-7, 25,15 EUR, 128 Seiten

Cuvillier Verlag
Internationaler wissenschaftlicher Fachverlag

Innovationen mit Mikrowellen und Licht
Forschungsberichte aus dem Ferdinand-Braun-Institut, Leibniz-Institut für Höchstfrequenztechnik

Herausgeber: Prof. Dr. G. Tränkle

Band 30: **Christian Fiebig**
Diodenlaser mit Trapezstruktur und hoher Brillanz für die Realisierung einer Frequenzkonversion auf einer mikro-optischen Bank
ISBN: 978-3-95404-690-4, 26,30 EUR, 140 Seiten

Band 31: **Viola Küller**
Versetzungsreduzierte AlN- und AlGaN-Schichten als Basis für UV LEDs
ISBN: 978-3-95404-741-3, 34,40 EUR, 164 Seiten

Band 32: **Daniel Jedrzejczyk**
Efficient frequency doubling of near-infrared diode lasers using quasi phase-matched waveguides
ISBN: 978-3-95404-958-5, 27,90 EUR, 134 Seiten

Band 33: **Sylvia Hagedorn**
Hybrid-Gasphasenepitaxie zur Herstellung von Aluminiumgalliumnitrid
ISBN: 978-3-95404-985-1, 38,00 EUR, 176 Seiten

Band 34: **Alexander Kravets**
Advanced Silicon MMICs for mm-Wave Automotive Radar Front-Ends
ISBN: 978-3-95404-986-8, 31,90 EUR, 156 Seiten

Band 35: **David Feise**
Longitudinale Modenfilter für Kantanemitter im roten Spektralbereich
ISBN: 978-3-7369-9116-3, 39,20 EUR, 168 Seiten

Band 36: **Ksenia Nosaeva**
Indium phosphide HBT in thermally optimized periphery for applications up to 300GHZ
ISBN: 978-3-7369-287-0, 42,00 EUR, 154 Seiten

Band 37: **Muhammad Maruf Hossain**
Signal Generation for Millimeter Wave and THZ Applications in InP-DHBT and InP-on-BiCMOS Technologies
ISBN: 978-3-7369-9335-8, 35,60 EUR, 136 Seiten

Band 38: **Sirinpa Monayakul**
Development of Sub-mm Wave Flip-Chip Interconnect
ISBN: 978-3-7369-9410-2, 44,00 EUR, 146 Seiten

Band 39: **Moritz Brendel**
Charakterisierung und Optimierung von (Al, Ga) N-basierten UV-Photodetektoren
ISBN: 978-3-7369-9465-2, 49,90 EUR, 196 Seiten

Band 40: **Erdenetsetseg Luvsandamdin**
Development of micro-integrated diode lasers for precision quantum optics experiments in space
ISBN: 978-3-7369-9479-9, 39,00 EUR, 126 Seiten

Internationaler wissenschaftlicher Fachverlag

Innovationen mit Mikrowellen und Licht
Forschungsberichte aus dem Ferdinand-Braun-Institut, Leibniz-Institut für Höchstfrequenztechnik

Herausgeber: Prof. Dr. G. Tränkle

Band 41: **Thi Nghiem Vu**
Development and analysis of diode laser ns-MOPA systems for high peak power application
ISBN: 978-3-7369-9480-5, 38,80 EUR, 138 Seiten

Band 42: **Christian Bansleben**
Differentieller Mikrowellen-Leistungsoszillator für die Realisierung ultrakompakter Plasmaquellen in Matrixanordnung
ISBN: 978-3-7369-9530-7, 34,90 EUR, 136 Seiten

Band 43: **Martin Winterfeldt**
Investigation of slow-axis beam quality degradation in high-power broad area diode lasers
ISBN: 978-3-7369-9733-2, 39,90 EUR, 158 Seiten

Band 44: **Jonathan Decker**
Investigation of monolithically integrated spectral stabilization in high-brightness broad area diode lasers
ISBN: 978-3-7369-9798-1, 49,50 EUR, 174 Seiten

Band 45: **Andreea Cristina Andrei**
Untersuchung und Optimierung robuster und hochlinearer rauscharmer Verstärker in GaN-Technologie
ISBN: 978-3-7369-9810-0, 39,90 EUR, 148 Seiten

Band 46: **Peng Luo**
GaN HEMT Modeling Including Trapping Effects Based on Chalmers Model and Pulsed S-Parameter Measurements
ISBN: 978-3-7369-9906-0, 48,00 EUR, 160 Seiten

Band 47: **Jörg Jeschke**
Entwicklung von optisch pumpbaren UVC-Lasern auf AlGaN-Basis Chalmers Model and Pulsed S-Parameter Measurements
ISBN: 978-3-7369-9918-3, 44,90 EUR, 176 Seiten

Band 48: **Nikolai Wolff**
Wideband GaN Microwave Power Amplifiers with Class-G Supply Modulation
ISBN: 978-3-7369-9931-2, 44,90 EUR, 170 Seiten

Band 49: **Carlo Frevert**
Optimization of broad-area GaAs diode lasers for high powers and high efficiences in the temperature range 200-220 K
ISBN: 978-3-7369-9944-2, 44,90 EUR, 174 Seiten

Band 50: **Bassem Arar**
GaAs-based components for photonic integrated circuits
ISBN: 978-3-7369-9976-3, 43,60 EUR, 152 Seiten

Cuvillier Verlag
Internationaler wissenschaftlicher Fachverlag

Innovationen mit Mikrowellen und Licht
Forschungsberichte aus dem Ferdinand-Braun-Institut, Leibniz-Institut für Höchstfrequenztechnik

Herausgeber: Prof. Dr. G. Tränkle

Band 51: **Mahmoud Tawfieq**
Development and characterisation of a diode laser based tunable high-power MOPA system
ISBN: 978-3-7369-9983-1, 44,90 EUR, 170 Seiten

Band 52: **Sebastian Preis**
Hocheffiziente frequenzagile Mikrowellen-Leistungsverstärker auf Basis von Verbindungshalbleitern und Ferroelektrika
ISBN: 978-3-7369-7004-5, 54,00 EUR, 136 Seiten

Band 53: **Simon Fleischmann**
Materialaspekte der Hydridgasphasenepitaxie von Aluminiumgalliumnitrid
ISBN: 978-3-7369-7029-8, 41,80 EUR, 160 Seiten

Band 54: **Erhan Ersoy**
Optimierung von koplanaren GaN-MMIC-Leistungsverstärkern im X-Band
ISBN: 978-3-7369-7157-8, 39,90 EUR, 154 Seiten

Band 55: **Simon Rauch**
Lateral emission characteristics of high-power broad-area lasers subject to external optical feedback
ISBN: 978-3-7369-7168-5, 29,90 EUR, 112 Seiten

Band 56: **Frank Dittmar**
Untersuchung der Strahlgüte von brillanten Hochleistungs-Trapezlasern für den Wellenlängenbereich bei 808 nm
ISBN: 978-3-7369-7167-7, 44,90 EUR, 176 Seiten

Band 57: **Florian Hühn**
Flexibler Modulator und Digitalverstärker-MMIC für den energieeffizienten Betrieb einer digitalen Sendekette im GHz-Bereich
ISBN: 978-3-7369-7190-5, 29,90 EUR, 162 Seiten

Band 58: **Dimitri Stoppel**
Interconnection development for InP-HBT terahertz circuits
ISBN: 978-3-7369-7204-9, 44,90 EUR, 156 Seiten

Band 59: **Anissa Zeghuzi**
Analysis of Spatio-Temporal Phenomena in High-Brightness Diode Lasers using Numerical Simulations
ISBN: 978-3-7369-7289-6, 49,90 EUR, 176 Seiten

Band 60: **Tasmim Alam**
Spectroscopic Applications of Terahertz Quantum-Cascade Lasers
ISBN: 978-3-7369-7297-1, 39,90 EUR, 132 Seiten

Cuvillier Verlag
Internationaler wissenschaftlicher Fachverlag

Innovationen mit Mikrowellen und Licht
Forschungsberichte aus dem Ferdinand-Braun-Institut, Leibniz-Institut für Höchstfrequenztechnik

Herausgeber: Prof. Dr. G. Tränkle

Band 61: **Max Schiemangk**
Ein Lasersystem für Experimente mit Quantengasen unter Schwerelosigkeit
ISBN: 978-3-7369-7293-3, 49,90 EUR, 166 Seiten

Band 62: **Thorben Kaul**
Epitaxial Design Optimizations for Increased Efficiency in GaAs-Based High Power Diode Lasers
ISBN: 978-3-7369-7396-1, 38,90 EUR, 136 Seiten

Band 63: **Pietro della Casa**
Two Step MOPVE, in-situ etching and buried implantations: applications to the realization of GaAs laser diodes
ISBN: 978-3-7369-7397-8, 72,90 EUR, 250 Seiten

Band 64: **Heike Christopher**
A compact mode-locked diode laser system for high prescision frequency comparison experiments
ISBN: 978-3-7369-7399-2, 59,90 EUR, 206 Seiten

Band 65: **Rimma Zhytnytska**
Design von GaN Transistoren für leistungselektronische Anwendungen
ISBN: 978-3-7369-7413-5, 57,90 EUR, 182 Seiten

Band 66: **Sebastian Walde**
AIN base layers for UV LEDs
ISBN: 978-3-7369-7451-7, 44,90 EUR, 156 Seiten

Band 67: **Jan Ruschel**
Ursachen der stromgetriebenen Degradation von UV-LEDs
ISBN: 978-3-7369-7542-2, 49,90 EUR, 180 Seiten

Band 68: **Erik Freier**
Optimierung der Prozesstechnologie und Steigerung der Zuverlässungkeit und Lebensdauer von (InAlGa)N-basierten Halbleiterlaserdioden
ISBN: 978-3-7369-7604-7, 49,90 EUR, 178 Seiten

Band 69: **Norman Ruhnke**
A deep ultraviolet laser light source by frequency doubling of GaN based external cavity diode laser radiation
ISBN: 978-3-7369-7613-9, 39,90 EUR, 144 Seiten

Cuvillier Verlag
Internationaler wissenschaftlicher Fachverlag

www.ingramcontent.com/pod-product-compliance
Ingram Content Group UK Ltd.
Pitfield, Milton Keynes, MK11 3LW, UK
UKHW022000190726
13853UKWH00004B/1641

9 783736 976269